W9-AOJ-213

YOUR
HIDDEN
CREDENTIALS

The
Value of
Learning
Outside of
College

YOUR HIDDEN CREDENTIALS

Peter Smith, Ed.D.
Lieutenant Governor, State of Vermont

ACROPOLIS BOOKS LTD.
WASHINGTON, D.C.

ACROPOLIS BOOKS, LTD.
Alphons J. Hackl, Publisher
Colortone Building, 2400 17th St., N.W.
Washington, D.C. 20009

Printed in the United States of America by
COLORTONE PRESS
Creative Graphics, Inc.
Washington, D.C. 20009

Attention: Schools and Corporations
ACROPOLIS books are available at quantity discounts with bulk purchase for educational, business, or sales promotional use. For information, please write to:
SPECIAL SALES DEPARTMENT, ACROPOLIS BOOKS, LTD., 2400 17th ST., N.W., WASHINGTON, D.C. 20009

Are there Acropolis Books you want but cannot find in your local stores?
You can get any Acropolis book title in print. Simply send title and handling costs for each book desired. District of Columbia residents add applicable sales tax. Enclose check or money order only, no cash please, to:
    ACROPOLIS BOOKS, LTD.,
    2400 17th St., N.W.,
    Washington, D.C. 20009.

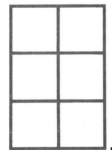

Dedication

I would like to dedicate this book to my wife Sally and my three sons, Benjamin, Daniel, and David. They all have been patient beyond the expectations of love, true learning partners in this personal learning project that their husband and father took on. It would not have been completed without their humor, support, and willingness to believe that there was something of value to be written.

Acknowledgements

I would like to thank the Carnegie Corporation of New York and the Fund For the Improvement of Postsecondary Education of the United States Department of Education for their grant support through the Mina Shaughnessy Fellowship Program, which supported my research for this book.

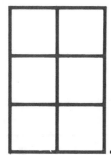

Table of Contents

3.

Successful Transitions 41

4.

On the Outside Looking In 61

5.

Adult-Friendly Colleges 81

6.

Bringing Two Worlds Together 97

Appendices 105

Bibliography 181

Introduction

I've been a personal learner all of my life. So have you. Personal learning is the knowledge you gain from your life experiences. It includes actual skills and abilities that are marketable, as well as the values, attitudes, and behaviors that you develop through reflection and introspection. John Dewey, a famous Vermonter, discussed the relationship between experience and personal learning in his book, *Experience and Education.*

> ". . . Every experience enacted and undergone modifies the one who acts and undergoes, . . . We often see persons . . . (who) have the precious gift of ability to learn from the experiences they have."

This type of learning is an untapped natural resource of extraordinary importance to a society that says that it values individual worth and wants to promote opportunity. Yet, most of us ignore it, educators, employers, political leaders, and individuals alike.

My first experience with personal learning came at the Colorado Outward Bound School in 1964. The Rocky Mountains were our laboratory and the curriculum was outdoor training and mountaineering. But we learned far more than the formal curriculum as we lived and worked in the wilderness of Colorado. I began to discover my own personal limits, the meaning of self-reliance, and the value of teamwork.

Then, as students of Princeton University, a group of us passed up the usual Saturday morning socializing to offer an educational program to local secondary school children. As we "taught" them, we learned about the reality of limited options and reduced futures for many poor and minority children who lived in the shadow of our great university. From that experience I developed the beginning of a lifelong commitment to redress the imbalance of opportunity in our society.

Since then, I have seen and used personal learning's value as a teaching tool several times. As a community organizer, I found that adults who did not read well improved their skills by helping nonreaders begin to read. Later on, as the architect of a high school community service internship program, I saw high school students learn helping skills as they worked with senior citizens in recreation programs. And, as the director of a street academy for dropouts, I saw law enforcement and the legal system come alive for my students as they observed lawyers and prosecutors at work.

Then, in 1970, I founded a small public college in Vermont that was dedicated to the lifelong learning of adults. Over the next eight years, we established a statewide program that evaluated personal learning done outside of college and awarded academic credit for it. To date, thousands of Vermont adults have used the program to establish their hidden credentials.

Initially, I saw our program as a practical attempt to avoid duplicating learning that our adult students had already done. But, beyond simply getting their credit, our students were profoundly influenced and strengthened by the recognition and rewards they received for their personal learning. The evaluation turned out to be more than just a good idea. It was a revolutionary tool for education and personal growth.

As I saw the changes that recognition of their personal learning caused in our students' academic performance and personal attitudes, I wanted to share the simple but powerful examples of their learning and life stories. Using several "portraits of learning" drawn from interviews with adult learners, I have written about personal learning as a universal but largely unrecognized influence on the human condition. *Your Hidden Credentials* is the result. I have written it in the belief that personal learning, if recognized societally, will have a revolutionary impact on our education and employment practices as well as our individual understanding of personal growth and development.

Using individual portraits of living and learning, *Your Hidden Credentials* describes personal learning, warns about the consequences of its denial, and promotes the power that comes with its recognition. The first chapter, "Personal Learning and Your Hidden Credentials," makes the case for recognizing personal learning and your hidden credentials in college and the workplace. The simple recognition exercises in Appendix 1 will help you identify some of your own personal learning.

Turning points are the times in your life when events demand that you learn, that you change your behavior, your skills, or your knowledge. Chapter 2 describes and analyzes several different kinds of turning points. It will help you recognize their pivotal role in promoting your personal learning.

Chapter 3, "Successful Transitions," is a series of stories about people who have used personal learning successfully in their lives. The planning exercises in Appendix 2 will help you identify the resources you have used to learn and plan consciously for more.

Chapter 4, "On the Outside Looking In," describes some of the problems and potential for growth that come from personal learning. These portraits also describe the costs of having the value of your learning denied by the outside world as well as the obstacles that most colleges and employers place in the way of adults. The exercises in Appendix 3 will help you draw meaning from the learning you have done and begin to identify your hidden credentials.

The fifth chapter, "Adult-Friendly Colleges™," describes how colleges can better meet the real learning needs of adults. Using the words of learners like you, the chapter tells what an adult-friendly education could be like. Appendix 4 contains a checklist you can use to evaluate whether or not a particular college or employer is adult-friendly. Appendix 5 lists those colleges that have indicated an interest in serving adults.

The final chapter, "Bringing Two Worlds Together," explores the historical and the current antecedents of the chasm that lies between the worlds of work and education. It includes a discussion of the forces that created this problem, an analysis of the consequences of our failure to cope with it, and recommendations for policies and programs that will close the gap. Clearly, if adults are going to be better treated educationally in order to be more productively integrated into the work force, this gap will have to be reduced.

In preparing *Your Hidden Credentials,* I interviewed over sixty people across the country—from Vermont to California—people who had grappled with this problem, many of whom found ways to put their personal learning to use. Their stories describe the richness and diversity of personal learning in America today, an extraordinary mosaic of learning and change generated by the experience of living. But, they also describe the frustration, the anger, and the confusion generated by our failure as a society to recognize the value of that learning. As you read, you might find yourself wondering if you are someone who is trapped by being unable to use your experience and personal learning. By the time you finish reading, my hope is that the persons featured here will have inspired you to consider your own experience and learning, to seek the opportunities that our folklore has promised.

1.

Personal Learning and Your Hidden Credentials

Your hidden credentials are the sum total of knowledge, behavior, and skill that has been generated by your personal learning. Personal learning has two aspects. It can result in the actual acquisition of concrete skills, an improvement in your competence. And, it can cause subtle changes in your attitudes and behavior, your sense of awareness of your changing self. If you are like most people, you haven't considered seriously the possibility that you possess hidden credentials, let alone tried to put a value on them. But, they are there in your life, an iceberg of knowledge showing tips of experience—significant events—with the larger body of knowledge concealed below the surface.

Identifying your hidden credentials depends upon you recognizing the larger, hidden meaning of your personal learning and placing an accurate value on it. As developmentalist Dr. Rita Weathersby has observed,

> "[People grow] . . . through an active process of making meaning from experience . . . Development apparently stops when people . . . have the experience but choose not

to use it to change their basic way of experiencing the world."

If you do not understand and value your experience and the learning it contains, you will be imprisoned by it, unconscious of its influence on your world view, your skills, and your knowledge. But as you come to terms with the personal changes that the accumulation of experience inevitably brings, you will go through transitions of age and development. Gail Sheehy called them "passages." If you lose track of your personal learning and do not value it, you will also lose track of who you are becoming and why. Then, your transition may well become a crisis of rediscovery.

But, if you recognize your personal learning, your transitions do not have to be crises. If you keep track of your learning you can begin to affect your personal development in a conscious, deliberate way by recognizing and integrating changes as they occur. Then, instead of being a crisis, the transition can be a reaffirmation of your most recent learning and your capacity to learn in the future.

Hidden Credentials Ignored

Our failure as individuals to recognize personal learning and hidden credentials is not unique. Colleges and employers do not value them either. At the work site and in the classroom, your hidden credentials represent a valuable accumulation of knowledge, skills, and behavior. But, for most colleges and employers, the tip of the iceberg is made up of credits, degrees, and credentials. When they ignore your hidden credentials, they are ignoring the bulk of what you know as well as the attributes that will make you successful as a learner and a worker.

Just how much personal learning you regularly accomplish is suggested by Allen Tough in *The Adult's Learning Projects:*

"Almost everyone undertakes at least one or two major learning projects a year. Some people undertake as many as fifteen or twenty. The median is eight projects a year lasting a total of eight hundred hours (15 hours a week).

Highly deliberate learning efforts take place all around you. The members of your family, your neighbors, colleagues, and acquaintances probably initiate and complete several learning efforts each year . . . When asked about their learning efforts, many of our interviewees recalled none at first, but as the interview proceeded, they recalled several recent efforts to learn."

The Scarecrow and the Real Need for Recognition

If you find yourself with no way of identifying what you know and no way of giving it a value in terms of your personal development, the marketplace, or higher education, remember the Scarecrow in the Wizard of Oz. Although he showed great ingenuity and intelligence throughout his travels with Dorothy, he believed that he knew nothing, that he "had no brain." So, he asked the Wizard for a brain. But, the Wizard knew better.

"Can't you give me brains?" asked the Scarecrow.

"You don't need them. You are learning something every day. A baby has brains, but it doesn't know much. Experience is the only thing that brings knowledge, and the longer you are on earth the more experience you are sure to get."

"That may be all true," said the Scarecrow, "but I shall be unhappy unless you give me brains."

The false Wizard looked at him carefully. "You don't need brains, you need confidence. Back where I come from, we have universities, seats of great learning, where men go to become great thinkers. And, when they come out they think deep thoughts, and with no more brains than you have. But they have one thing you haven't got, a diploma. Therefore, by virtue of the authority vested in me by the Universitatus Comitiatus E Pluribus Unum, I hereby confer upon you the Honorary Degree of Th.D.—Doctor of Thinkology. . . .

"Oh!" said the Scarecrow. "The sum of the square roots of any two sides of an isosceles triangle is equal to the square root of the remaining side. . . . Oh joy! Rapture! I've got a brain."

The Scarecrow thought he needed brains. But, what he really wanted was acknowledgment and recognition that he was a knowledgeable person with skills and learning that were valuable in the eyes of the world and to the people around him. The Wizard's recognition of what he knew gave the Scarecrow the confidence he had been lacking. With that recognition and the confidence it inspired, the Scarecrow was empowered.

You are in a similar situation with colleges and employers. When they don't recognize your hidden credentials they are denying you the benefits and the power that would naturally flow from them. This is a wasteful and destructive practice in several respects because:

- It buries the value of your experience and loses your learning to all who would benefit.

- It demeans adults who have significant personal learning and who want to benefit from its worth.

- It perpetuates the connections between self-esteem and confidence on the one hand and formal educational attainment on the other.

- It transforms college from a learning experience into a hunt for the credential, "that piece of paper."

This explains the angry reaction I received when I met with a group of child-care workers and parents, deep in the hill country of northern Vermont called the Northeast Kingdom. They wanted a college education, and I had just finished describing a new college for poor and working Vermonters designed especially to bring education to them in places and at times that were convenient. I had explained carefully that we would not be tied to academic tradition. There would be no credits or degrees, just courses that interested them, like child development.

But one of the women there, Margery Hood, wasn't buying it. There was anger in the air, averted eyes, heavy silence, shifting chairs.

"You, with your college degree on the wall, are telling us we don't need one!" Her face flushed and voice rising, Margery continued. "What do you think we are, dumb? We *need* credit. If you're not going to be tied to college traditions, you should start by giving us credit for the things we have already learned working here."

Margery needed help all right, but not the help I was offering. She was trapped by a reality that I had ignored. Everything she knew she had learned raising her family on a Vermont farm and working in the child-care center. She was very good at her work. But, colleges and other employers did not recognize or accept that kind of learning when she looked for another job or tried to further her education.

As she saw it, the system had no room for her or her learning. Credits and degrees were tickets to a better life. But, since colleges give credit for what they teach and employers look for that credit when they hire, a program that didn't recognize the learning she had done outside of college and didn't give any credit would have little value to her. Margery would have been wasting her time at my college.

As I drove home that night, I questioned myself.

- Should child-care workers with years of experience on the job as well as job-related training be treated like 18-year-old freshmen by the college?

No.

- Did it make good educational sense to recognize and reward learning that had happened away from school and give credit for that learning as well as the coursework offered?

Yes, if we did it well.

- Should an employer look beyond paper credentials to a person's actual experience and ability when assessing their potential and capacity to perform?

Yes, if they wanted to get the best that each person had to offer.

We needed to bridge the gap between the personal lives of our students and their college program, blending the learning they had done informally with courses that took them further. There had to be formal recognition of their informal learning so it would be useful to them in the outside world.

Because of Margery's "advice," our college, a small experimental school for low-income, rural adults, decided to look at new students to see just what knowledge they brought with them to school. It made no sense to me, nor to the men and women who became my colleagues, to require that students duplicate learning they had already achieved outside school. Rather, we reasoned, it made far better sense to figure out how to recognize and credit that learning as part of their college program.

This new approach unleashed intense and conflicting forces on the college and our learners. Other colleges, opposed to our policy and fearful of a loss of power and control, achieved an effective blockade of our students. Credits and degrees were suspect, transfers denied, employers suspicious. "Colleges provide instruction," the argument went. "You can't give credit for unsupervised learning that happened away from the college and before the student enrolled. The degree should stand for what we teach them in college, not what they have learned someplace else." Some in the business community wanted to know how they could be sure the learning was "real" if it hadn't happened in a college program.

But, our adult students' reactions were a stunning contrast to this academic embargo. One woman's response typified all of them: "Thanks for helping me see how much I had learned outside of school, how much it was worth, and what kind of a person I have become. Now I know I'm a learner and I'll never stop learning."

Our students found that they had learned a great deal during their lives, such as the man who had learned how to speak publicly and how to use a computer as the head of a local volunteer agency. But, they had devalued and "forgotten" most of their personal learning. The college's recognition of that learning gave them acknowledgement and an academic base to build upon. In several cases, they had learned virtually the equivalent of a college education without receiving any credit or taking any courses. I remember one woman, for instance, who had managed her own small business for ten years, learning from her experience, her peers, and the reading that she had done throughout.

Coming to see this learning and its value filled these individuals with pride, a sense of worth, and confidence. They had a bridge between their personal learning, its meaning, and its worth in the world. And, a large majority of them went on to further success in other schools and in the marketplace. Our college program was very successful.

Margery Hood was like most adults. She was trapped by an American contradiction that denies some of our most basic beliefs about learning, success, college, and jobs. The contradiction is simple, but its effects are complex and devastating. On the one hand, we have our historic belief in opportunity, in success for the individual. Anyone can achieve anything he or she wants to. On the other hand, we have colleges, universities, and employers demanding specific credentials before allowing people to progress. And most damaging, people—as a result—do not recognize their own skills and abilities acquired through work and other activities. The path to success is blocked for many people.

The American Dream and Personal Learning

American folklore is filled with the mythology of opportunity. This mythology, borne by phrases and stories, is the keystone to the American dream, delivering its message again and again:

Live and Learn.

The School of Hard Knocks.

Horatio Alger.

Rags to Riches.

In each case, the message is similar. You can learn from the experience of living. There are strong and useful connections between effort, learning, and success. We believe deeply that, if you are resourceful, work hard, and learn on your feet, opportunity will come knocking.

Part of the mythology is correct. Most people do learn actively and purposefully all the time. In fact, as Allen Tough found, the median adult spends over 15 hours every week on highly specific learning projects. Colleges account for only a tiny part of all that learning. Even if you don't consider it to be learning, you are learning when:

- you read about child development and discuss parenting with a friend or in an organized group in order to cope better with the children in your life; and when

- your supervisor at work demonstrates a new technique or the company brings in a lecturer to describe a new development in the industry; and when

- you develop a health and diet plan to keep physically fit while eating well; and when

- you read several books on China or Mexico to know more about those countries or to plan a trip; and when

- you study investment practices and then practice with your investments in order to make more with the small amount you have laid aside; and when

- you discuss and read about soils and seeds prior to starting your first serious garden; and when
- you develop better ways to manage your household and your finances; and when
- you learn to play an instrument, engage in therapy, or learn a new sport.

This learning has three important characteristics:

It is personal. You usually learn alone or with one or two other people.

It is purposeful. You always learn for a reason, even if you forget the reason or consider the learning unimportant. As a result, much of your learning is absorbed into your collective experience and forgotten.

It is powerful. Personal learning continually changes you, developing your behavior, skills, knowledge, and attitudes into a unique set of hidden credentials. But, you rarely think about what you've learned or how you've changed.

So, when our mythology tells us that we learn from experience, it is telling the truth. Unfortunately, as Margery Hood knew, the mythology breaks down when it promises a connection between personal learning and success. Formal education and our credentialed society have made this promise far less certain.

Most of us have heard about the stories of Horatio Alger, typically of a poor orphan who made his way in the world and became successful through hard work and pluck. They were success stories that caught the imagination and hopes of generations of Americans as a symbol of what was possible through hard work and effort. The sad truth is, if Horatio Alger's hero attempted to succeed in the same way today, he probably would be trapped in a dead-end job, caught between an education system and a credentialing apparatus that neutralized his efforts and rendered his native ability useless.

Although our folklore says that we believe in the value of personal learning, most colleges simply ignore experience and personal learning as they prepare students for a set of life opportunities that require credits, certificates, and degrees. The mother's experience with her children, the soldier's training as a commander, or the Peace Corps volunteer's exposure to a foreign culture are largely ignored as learning and as a base to build upon. Often, you can transfer credits from one accredited school to another, but you can't transfer learning from life. And, in the world of work, it is the rare employer who will look beyond academic credentials to other evidence of skills and knowledge when a new job opens up or a promotion is being considered.

The same system of rewards and credentials denies you mobility in the workplace.

- A woman returning to work outside of the home after years of raising a family, managing a household, and organizing hundreds of community events finds there is no way to cash in on the knowledge and skills she has developed over the years.
- The senior employee is forced to train her new supervisor for the job she can do but can't have because she doesn't have the degree.

It is a system that, at its worst, is dominated by the academic marketplace and emphasizes credentials over competence, overlooking the very human resources it is supposed to strengthen. In the process, we lose an enormous amount of talent and human potential. But, if your personal learning is denied—by you, a college, or an employer—you lose control of a precious resource. You can be trapped: knowledgable with no way to respect your learning, capable but unable to explain the capacity, changed but unable to describe the change; a prisoner of your experience.

As the next chapter shows, personal learning brings problems as well as the potential for growth. You are continually expanding your hidden credentials, taking in experience and information, absorbing their worth, and expelling the rest. The learning you do is as inevitable, yet varied, as the seasons of the year.

Turning points are those moments when you acknowledge that you have changed, having gained skills and new attitudes. Turning points can be times of wrenching change, but also enormous personal growth as you take stock of the personal learning you have done and discover the person you have become.

2.

Turning Points

Whether or not most people are aware of the personal learning they have done, at some point we all recognize that we have changed over the years. This recognition can range from a casual observation of the characteristics of aging to a more profound understanding of growth and development. For instance, the common aging process reminds us physically that we are changing, such as with the appearance of more grey hair. Many of us recognize that change has occurred only through such casual observations.

Although turning points involving personal learning are more profound, they are equally natural consequences of the same aging process. As I stated in Chapter 1, you continuously acquire knowledge through personal learning as you grow older. When you acknowledge the changes in your life and the personal learning you have accomplished over time, you bring yourself to a turning point. Turning points can mean the most to you, regarding your personal learning, if they initiate the following progression of events.

- You realize that you have to change, that things can't go on the way they are. You reappraise yourself.

- You gain confidence and a sense of self-worth as you take stock of your personal learning and become aware of the skills, abilities, and values that you have gained over the years.

- And, with your increased ability to reflect on your learning, you become a conscious learner, anticipating future turning points and enhancing the aging process with continuous and purposeful learning.

Although your personal learning actually causes personal change, turning points, the moments when you recognize that change, are often triggered by crises in your life: a lost job, a divorce, or an illness in your family. The key question, though, is how you react to a turning point. Will you simply fail to grasp it? Will it overwhelm you? Will it be a wrenching experience that leaves you looking at a stranger in the mirror wondering who you have become? Or, can you convert it into a moment of learning and self-realization that changes the way you see the world, stimulating more learning and augmenting what you already know?

Significant turning points hinge on a sudden realization of self-worth that comes with the recognition of knowledge that has been learned personally over the years. Sometimes the learning exposed at a turning point is personal, involving your behavior, values, and relationships with other people. At other times it is career-based, relating to your knowledge, skills, abilities, and opportunities. Either way, a turning point represents a strong change in your perception of yourself in the world and a realization of your self-worth.

Unfortunately, most of us reach our turning points without the ability to control or understand the forces they unlock. Generated by outside events that we apparently can't control, turning points often either pass us by completely or cause wrenching change in our lives.

Elaine McDermott

Elaine McDermott of Montpelier, Vermont, was a math teacher before she raised her two daughters. She had always planned to return to teaching when her children were in school. However, a confusing and agonizing turning point stimulated new insights and led Elaine to make a very different decision about her career.

". . . All of a sudden I said, "Wait a minute. That's not me. I don't think I'm going to do that. No, I'm not going to do that."

Your Hidden Credentials

For years I had been thinking that when the girls were in school, I would go back to work teaching. That first year I took a temporary job at a local coin company running their retail store. I was planning to apply for teaching jobs for the following fall. My whole attitude about running the store was that this was fun. I would try it for a while. I knew I wasn't growing. It was just bucks and it was boring, but it was nice to get out, and I had an excuse to buy clothes. I kept telling people that I was just doing it for now, until I could get a job in June teaching, because I wasn't growing, I wasn't learning anything. I felt that I had control over it. Then, they took it all away from me, on New Year's Eve.

There had been two major layoffs before then so that the writing was on the wall. But, I still thought that I had a couple of months, if it was going to be their choice. If it was going to be mine, I had until June. I couldn't foresee that my job was going to end in December.

I can't identify the feeling that I felt at the time. We went out and had drinks and laughed, but it was just . . . there was a strange atmosphere because you were making believe you were happy about something when you really weren't. I was really going to miss that money, and it called my bluff about other jobs.

I was in a sort of state of shock for three weeks. It sounds like a strong statement to make, but I think I was, because, as I look back on it, I couldn't identify what was going on at the time. I was staying home saying, "Isn't this just wonderful, I can spend all day reading." But it was awful, and I couldn't recognize that. After about a month, I identified the anger, which was good. But then, the depression hit, and I thought, "My God, what am I going to do? My kids are in school, and I'm 35 years old, and I've got to get my act together."

I have a friend who is going through a very difficult time in her life, a divorce—not by her choice. She is the victim and she's going through a lot of things that a lot of women go through who have followed their husbands around as their careers and haven't established anything for themselves. I thought about what she was going through, and I thought that if I didn't find something, if I didn't get my act together, I could be that same woman.

So I said, "Okay, you can't wait until June anymore. Now, what are you going to do?" Well, I was going to teach because that had been my pat answer for the past ten years. So, I started to send resumes to different area schools. Then I started getting replies. That was the thing that threw me into a depression, because, at that point I had to make a commitment. Writing the letters was no commitment.

I realized that I didn't want to teach anymore. It was just something that I hadn't thought about for ten years. I always felt that that's what I would do when the kids were grown. Plus, George was encouraging me, too. Which he would do because he'd been listening to me for ten years, and here I was, applying for these school jobs, so why shouldn't he think that? But, I heard him on the phone one day, telling his parents that I was going to teach, and, hearing it like a third party, all of a sudden I said, "Wait a minute. That's not me. I don't think I'm going to do that. No, I'm not going to do that.

During those years at home, Elaine's life changed outwardly. But she wasn't aware of a concurrent inner change in her personal aspirations and expectations. They had changed so slowly that she continued to carry a clear but outdated image of herself as a teacher when she returned to work. It wasn't until circumstances gave her no alternative that Elaine confronted the changes in herself. She became more aware of her values. She didn't want to teach. That's when she reached her turning point.

It took Elaine ten years to reach her turning point, even though she had been changing the entire time. A deeper self-awareness was necessary to complete the changes when the turning point came. Only then was she primed to seek further personal learning by training to be a paralegal, thus adding to her new self-awareness brought about by her turning point.

I had always been interested in the law. At one point, I had regretted that I hadn't done something in law when I got out of college. I had heard of the school, Woodbury Associates before, but I just kind of filed it. Every now and then, I'd bring it up with people when I still thought I had a choice. But it was just part of something I was faintly considering.

I was just not knowing what I was going to do. There was an orientation meeting the following Thursday. I thought, "Maybe that's what I need, right now." So, I gave them a call and told them I'd be there. That just unlocked something. It cemented the decision and my resolve to go. I felt as if I had made a commitment to do something that had been in me for a long time. I sat down that day and wrote two letters, one to my in-laws and one to a friend, and I told them that I was going to a paralegal school."

As difficult as it was, Elaine seized her turning point and put the learning it generated to work for her. Elaine learned self-awareness and gained insight into her motives and aspirations when she finally confronted the consequences of being laid off. Then, she grappled with the powerlessness that comes from a lack of direction and chose a different course for her life.

Joe Lamell, on the other hand, had many of the skills and the self-awareness that he needed for a new direction in life. But, he missed several turning points because he wasn't ready to change his behavior and values.

Joe Lamell

Joe Lamell grew up tough, "a punk" in Montpelier, Vermont. He had been "everybody else's fall guy" for years. As we sat at his kitchen table, he admitted to having been swept along by life. But, after he lost his job working at a local college because of a service-related injury, Joe realized that "riding on someone else's deal" meant being a victim of life's circumstances.

"It made me stop, open my eyes, and say, 'Look, this is what happened to you when you were in the service, and now you are getting the short end of the stick up here. Now, that's the way it's going to be for the rest of your life if you don't change it. So, let's put the brakes on and take a different route.' "

I grew up in Montpelier. I was a smart-aleck wild punk, the leader of the Court Street gang. I had about ten followers and, whenever there was trouble or anything else, it was "Go get Joe. Let's go." You know? That was me. I was on my own at the age of 15 and I have had to get and do a lot of things on my own ever since. That's going on 18 years now. I spent ten years in the Marine Corps, three tours in Vietnam, before I got out in 1975. I've been married three years now, working off-and-on jobs, and going to school. The main reason I'm trying to further my education is the fact that I'm tired of being everybody else's fall guy, or dope guy.

A lot of my attitude came out of the service, especially my last two years, when they sent me down to Quantico and I didn't want to go there. I didn't want to be around them 18-year-old college punks. Seeing them get 2nd Lieutenant bars and they are going to turn around and tell me what to do after I have had eight years . . . I've trained these people, I did three tours in 'Nam. I didn't think I was God, but I knew what I was doing, and I didn't feel that, with eight or nine years in, that somebody coming out of boot camp with a bar on was going to tell me what I was supposed to be doing.

A lot of the difference was that they were book-learned, book-taught. They had it upstairs and they figured they could go out and assault a hill using the book. And, you know, you just didn't do all things that way. I mean, it's great to be able to teach certain things by the book, but when it comes to actual combat and preparing a person for a combat situation, you have got to teach them the know-how, the basic idea of staying alive. It comes in awful handy and an awful lot of men over there in Vietnam never saw the end of the first month because they were so smart.

My knowledge was actual living and learning, not book knowledge. I didn't get any of my knowledge out of a book on how to dodge a bullet or how to cover up when you are hit by mortars. Mine was all actual true-life living. I think that's the biggest difference right there.

Two years ago, during deer season, I was working at the college and I hurt my leg. It was a recurrence of an injury from Vietnam and I was laid up for six or seven weeks. Then, the doctor told me that they had to operate. So, I went back to the university and told my boss what was happening and he says, "Well, if you're not back here by December 14th, we no longer need you." Now, I had doctor's papers and everything stating that this was service-related, and I was going to be operated on, and that I was not capable of working. Still, they let me go. I was operated on the Friday before Christmas, 1979. Right then and there

I figured to myself, "Yeah. That's the kind of screwing I got." And I said, "No more. That's it."

I felt, "By God, I did three tours in Vietnam. I served when, you know, it was tough." Now, I was not capable of working and they still let me go. Then, I looked at my pay check and I said, "Christ, I worked forty hours for $130.00. This ain't making a program. I've got a wife and three kids that I've got to handle."

I'm not going to say that I was 100 percent content before that, but I was willing to ride on somebody else's deal provided I didn't wind up getting the short end of the stick. Then, at the university it just put through the point. It made me stop, open my eyes, and say, "Look, this is what happened to you when you were in the service and now you are getting the short end of the stick up here. Now, that's the way it is going to be for the rest of your life if you don't change it. So, let's put the brakes on and take a different route.

Joe missed a major turning point when he reacted negatively to his transfer to Quantico. At the moment of crisis he possessed the personal learning of self-awareness. He believed he knew more about soldiering than officers fresh out of training school. And, his past learning about soldiering in Vietnam supported his sense of self-worth, that his knowledge was more valuable than learning by the book.

But he didn't do anything about it and the opportunity presented by the turning point turned into more frustration. He had neglected an important body of past personal learning and interest that contributed heavily to a more successful turning point later on.

I think a lot of it is my attitude. It has changed considerably. I have been able to do a lot of talking, getting a lot of things out, whereas a year ago you couldn't get me to talk. I thought I was the only one carrying around as big a load of these problems, service-connected. Now, I know I'm not the only one.

I want my own place in the future. I want to make the system work for me instead of me working for the system. They are going to work for me.

My first three years in the army I was a grunt rifleman. I couldn't foresee any civilian job on the outside that being a rifleman would qualify me for, except maybe a professional killer or something. So I got interested in food services, the food service field. I reenlisted for it, and they sent me to special Mess training. When I finally got out of the service, I went up to Stowe and went to chef's school at the Toll House.

Then, after I lost my job at the college, I went to the Voc Rehab at the VA and I was going to get into aircraft mechanics. But, due to the fact that my leg was injured, the counselor pointed me towards food services again. He said, "You've got 15-16 years in food services, why don't you just go that route. Why don't you just stay in the food services field and shift it around a little bit and go into management?"

I thought that I had hit my peak in food services, but that was just my idea. There was a lot of learning I got out of those experiences. I got some of the basics. I had managed before and I had been an assistant manager, too. Now, I'm trying to take what I know and put it to work for me.

In the next five years or so I want to be able to own and operate my own restaurant or my own facility. It is becoming more realistic, now, because of the fact that I'm getting the training, and I'm getting a broader spectrum of what needs to happen and how its got to be done. The basic thing is, I have got my food service knowledge behind me. Now I'm getting the basics, like bookkeeping and management. I honestly feel that, with my knowledge, my actual working knowledge, plus combining it with my book knowledge, it is going to make it just that much better for me.

Joe felt at odds with the world around him for a long time before he capitalized on one of his turning points. When the university let him go, he faced another internal crisis. He felt that he had been laid off unfairly because of a chronic disabling injury. But this time Joe seized upon this critical turning point, and converted it into an opportunity. He drew upon a body of knowledge that he had previously ignored or forgotten, his food service experience. When he went after additional personal learning in this area, Joe took the ultimate step toward a successful turning point: he added new information to his existing body of knowledge that presented him with new prospects and a chance to deal with future turning points in similarly satisfying ways. Joe had finally combined his own sense of self-worth and past skills with

Your Hidden Credentials

a willingness to broaden both, so that he wouldn't miss any more of the opportunities that accompany turning points.

We've seen that both Joe and Elaine failed to capitalize on some of their earlier turning points even when they recognized the need to change. By recognizing this need, they experienced the first crucial stage of converting a turning point into a significant opportunity. But, they each progressed only part way in following through on the second step, that of taking stock of their personal learning.

As I mentioned in Chapter 1, personal learning possesses two aspects, that of a body of skills previously unrecognized and unacknowledged by yourself, and also the awareness of your changed inner self, including new values and a sense of self-worth. Elaine neglected to see her changed self, particularly in the area of the kind of work she wanted to pursue. Joe knew his self-worth, but he'd forgotten the other aspect of his personal learning, his skills in food service. Since both Elaine and Joe had not identified all of the aspects of their personal learning, they were unable to go beyond the recognition of a number of turning points until they came up against turning points that they could not ignore, wrenching turns of events that catapulted them into change without any preparation.

Martha Lucenti's story demonstrates this situation most dramatically. Martha never had identified her own sense of self-worth or her personally learned skills until a calamitous personal crisis thrust her into a situation in which change became a matter of personal survival.

Martha Lucenti

Martha Lucenti twisted her handkerchief throughout our conversation. After being divorced by her husband in the early 1970's, she went back to work and also began to attend Governor's State University in Park Forest South, Illinois. There was determination in her voice as we talked in the school's student lounge on a sunny day.

As Martha described the crisis of her "disastrous" divorce, she painted the picture of a person in utter despair. Her divorce led her to a potentially catastrophic turning point. Alone, with no sense of her own worth and believing that she was a "nobody,"

Martha confronted the choice between giving up, surrendering to the crisis, and avoiding the turning point, or trying to learn from it and reconstruct her life. In this elemental struggle, Martha discovered a gutsy resolve in herself. Then, beginning with her values and her behavior, she began to learn from the unanticipated trauma that her turning point had generated.

"Then, all of a sudden, you start to realize there is more to life than just becoming stagnant and doing the same things over and over. You're always seeking whether it's a new craft item . . . or a new way to bake a cake . . . or a new book to read about a new subject. I find that as I learn more, the more I see there is to learn."

I had a rather disastrous time in my life when I went through a divorce and all of the problems that an ex-husband can make for you and your children. The world kind of blew up in my face. One of the biggest problems was that I had spent so much time making my family think that they were special and they were something, that I never found time to make *me* be something or *me* be special. The stressful period I was going through was actually me thinking, "I'm a nobody, I'm a nothing. I have nothing to offer life. What did I do wrong?" I took the very grim side of the picture.

Instead of giving up, one day I woke up and said, "Now, wait a minute. I am somebody, and I'm going to prove to myself and the world that I am somebody and I can amount to something. I am of value." I know you can get a very low opinion of yourself when you are going through a stressful time; I refused to let that take over.

Now, I am projecting myself as an individual more so than trying to project my family and the children and my husband as being the great ones. I'm not just there as the carpet at the front door that you walk over each time you come in.

As she gained confidence and some personal momentum, Martha experienced the three kinds of learning that are a must if a turning point is going to be successfully converted. She learned about her values and skills and became more self-aware as she reflected on her life to date. Most importantly, she was able to move beyond reflection and self-evaluation and use the turning point and the crisis that caused it as the basis for further personal learning.

Many people destroy themselves during terrible times of stress. I actually used going to school and working to avoid that. I was trying to create and build on what I already had. And, by keeping extremely busy and involved, I went through one of the worst crises in my life and came out with flying colors. I'm very proud of myself. I tried to do something creative rather than destructive at a time of stress and it was the best medicine going.

Now, I'm remarried and I have a husband who is most supportive of what I'm doing. I have gotten involved in political movements here at the university and outside of the university. I have gotten involved in village politics. I've become more aware of what the world has to offer and what I can get.

I always used to think that I was in this little cocoon; that nothing could ever happen to me. Then, when something did happen, I was high and dry. Where do you go? What do you do? You have to learn how to get credit. You have to learn how to start your savings all over again and how to build up your estate for whatever you are going to leave to whomever you're going to leave it. You become very aware of the things that are available to you as far as people that you are going to meet, places you are going to go, and contacts you are going to make throughout the rest of your life. Those are all learning experiences.

I think you get to a point in your life where you think, "What's its value? Why do I have to keep learning? Haven't I learned enough sufficient to keep going?" Then, you reach a plateau where you are just kind of satisfied with what you do know.

Then, all of a sudden, you start to realize that there is more to life than just becoming stagnant and doing the same things over and over. You're always seeking, whether it's a new craft to learn, or a new way to bake a cake, or a new book to read, about a new subject. I find that, as I learn more, the more I see there is to learn. I don't think you can ever learn too much.

Martha's divorce forced her to a crossroads. Her turning point, as unanticipated and excruciating as it was, gave her the opportunity to learn and to change. Using her self-awareness and her personal learning, she seized this opportunity sensing that both her self-esteem and her identity were at stake.

We've discovered through these stories that turning points can be missed moments of opportunity if we ignore any of the stages in developing our personal learning. Fortunately for Elaine, Joe, and Martha, all of them eventually learned to make the most of critical turning points in their lives, even though they missed earlier ones. In fact, between their various turning points, they each learned something about themselves that helped them change for the better. Even this simple observation shows that people learn throughout life. Elaine, Joe, and Martha ultimately recognized their important turning points and the need for change, comprehensively took stock of their skills and values, and went on to pursue further learning to supplement their personal knowledge and achieve their desired goals.

Connie Yu Naylor's story, however, sheds more light on the dynamics of personal learning and turning points in life. Connie found out that capitalizing on one turning point doesn't signal the end of the need to learn.

Connie Yu Naylor

When I spoke with her, Connie Yu Naylor had already seen the world. Before becoming a commodities buyer for Ralston-Purina in St. Louis, she had worked on three continents in six jobs over a fifteen-year period. But, in her own terms, as she moved around, Connie had felt . . . "like a chip of wood carried on the water. Not very smart, a hard worker."

"I'm surprised at all the things that I have learned that I wasn't thinking about. I didn't realize I wanted to learn American philosophy because I was lonely. I thought it was just, you know, a seeking of knowledge. But there's always a reason why you learn, isn't there?"

I was born in Amoy, China, in 1936, just before the outbreak of the Second World War. When I was three, we moved to

Hong Kong. During the Japanese occupation, I had a mixed childhood with erratic schooling. I never did complete any formal kind of junior school education. I went to Japanese school for a year and Chinese school for a year and when the whole thing settled down, I was 13.

I don't know why, but I decided that I was too cloistered, and the only way to break out of that world was to get a job that included travel. So, after high school and over the objections of my parents, I became an air hostess. For three years, I traveled around the world as a BOAC air hostess. Through traveling, I began to know that there are other cultures in this world, not just the Hong Kong culture, which is cosmopolitan, but very small-town. And, I began to recognize a lack of polish in myself. Like being able to think and recognize art and literature. And, to understand that the world isn't just a Jane Eyre book. There's a lot more to it than that.

Finally, I went back to Hong Kong to hunt for a meaningful job. I worked first at a training school for Pan American where they taught people how to use computers to log passengers, cargo, and all that. They had some management seminars and I met a few Americans from good families with good educations. I began to see what was possible, but I always thought it was possible for them and not for me because I had passed that point where I was supposed to be in school.

So, I knew that I'd have to gain my experience through work. After three years at Pan Am I found an American company that was hiring people to sell municipal funds— International Overseas Investors. Again, that took me out of Hong Kong. By then, I liked traveling, and the job offered a very lucrative salary in a high-roller world. Traveling is a way to learn, so I started traveling again to Europe, to Holland, to Amsterdam.

I did a lot of learning with people who were in the mutual funds business, but they were a strange breed. Some were college graduates. But there were some who had never been to school, and they operated very differently. It was mostly high-pressure salesmanship. What we did was start a sales operation, hire Dutch people, and have them go out and sell. We were living high off the hog, lots of jewelry and fast cars.

But, something happened and the whole thing came crashing down. The company broke up and I went back to Hong Kong. That was my second shock. The first shock had been going from a very cloistered home into this high-flying world over a six-year period. The second shock was almost overnight. I ran. I went home and I was out of it.

As she was swept along during her early years, Connie did capitalize on her first turning point. But when her "high-flying" world working in Europe in the mutual funds business came crashing down, Connie ran from it, both physically and behaviorally, back to the cloistered world she had come from. Her personal learning was on hold during this period of time. After she bottomed out, Connie was unable to exploit her experience with people, cultures, investing, or banking. Instead, she went back to square one.

So, I was back in Hong Kong, with no education to speak of. Formally, I had high school, with an honors-type degree. Informally, I had a lot of background on how the rest of the world is. I was neither fish nor fowl. I had great expectations from the world, but nothing to achieve them with.

I went to a bank looking for a job and I must have been a good salesperson because the manager said, "I need an assistant and, even though you're not formally educated, you're it."

He was the chief manager of Barclay's Bank in Hong Kong. He knew how to bank without looking at anything and he recognized that I knew things but that I didn't have any formal education. He started me checking people who wanted loans. These were loans to build ships, railways, very big loans, upwards of 30 million pounds.

Even then, I was very aware that I didn't have the formal education to support what I was doing. What I did was by guesswork and by being shrewd. I knew there were ways of figuring the balance sheet—margin, ratio, and all that. But I knew that I had to go back to school to get the formal education, but I couldn't do it then. It was either keep my job, my income, or go back to school and lose them. I was stuck.

Then, I met a fellow, an American who was passing through. I fell in love with him and decided I was going to come here and visit him to see if he was really okay. I came,

Your Hidden Credentials

and I still liked him, so I applied for a visa to come and live in the United States. I moved to Boston in 1974.

We were married, and we moved to St. Louis. I had to quit my job and I thought, "Oh no, here I go again. I have nothing in my portfolio but experience."

I got a recommendation to Ralston-Purina. It was "Go and see these particular people and try and get a job." So, I went there and was given a good job as manager of a good department.

Now, I'm a commodities buyer. When I came in, I knew nothing about commodities, so that was another big change. I grabbed everything I could and read it. I listened to everyone I could, I talked to everybody and gradually learned how to buy for Ralston: things like meat, grains, and vegetables.

When I first came to America, I didn't really like myself at all because here I am, a strange creature in a country of people who all know what they're doing. And I began to realize that there were things which mattered to me that didn't seem to matter to most people around me. I was very lonely, and I had to decide either to try to reach people, to communicate, or go back where I belonged. Finally, I thought, "I'm here. This is my life. I must reach these people."

Well, first I tried to read people's minds. But how do you read people's minds without understanding what makes them feel, what made them become what they are? There are mental thought patterns which are different—your philosophy and history. So, I went back to American philosophy because I wanted to understand, to find out the American instinct. And I started reading many books on your history and culture, and on western interpretations of eastern meditation to see how your world tries to understand mine.

Now I'm becoming aware of the riches of the people around me. It's like discovering souls, lost dimensions in people. And, the more I'm beginning to understand and reach people, the more I am giving of myself. I think it flows together, like a link. I'm feeling better and better about myself.

And I'm surprised at all the things that I have learned that I wasn't thinking about. I didn't realize that I wanted to learn American philosophy because I was lonely. I thought it was just, you know, a seeking of knowledge. But there's always a reason why you learn, isn't there? Why was I feeling sorry for myself? I've done a lot. I'm also suddenly aware of where I am and what my expectations are. I have a long way to go.

Immersed in a new culture for the second time, the crisis of culture shock created another turning point for Connie. After enormous difficulty, she used her turning point to begin learning again. As Connie struggled to make sense out of the new culture she lived in and to improve her performance at work, she used her new personal learning to cope with the vast changes in her life. She increased her self-awareness and her values began to change. She decided that money was not as important as her own sense of well-being. And she learned actively and consciously about her new culture, western philosophy, and her job. As she learned, she began to understand the connections between her history, her ability, her behavior, and her daily environment.

Through this process, Connie developed extraordinary under-standing and introspection as well as the competence to do her job. She also adjusted her own self-image from being "just a silly female" to being a person who had "something to offer." Connie reappraised herself. She found capabilities she hadn't seen before as well as new values, an appreciation of her own intellectual ability, and a sense of how and what she had learned at work. Most importantly, Connie used her self-awareness and her knowl-edge to become a purposeful, ongoing learner. Connie made the transition, after her second turning point, from being a victim who was in conflict with the world to a person who was in charge of her personal learning and destiny.

> Over the last three years at Ralston, I have gradually come through a change in my self-evaluation. At first, I thought I was just a silly female, not very smart but a hard worker. Now people are beginning to listen to what I say. I'm begin-ning to realize, "Hey, I have something to offer"; that I am, you know, someone who can take my role and my part in this world. I'm suddenly comfortable with people whom I was very uncomfortable with before because I thought they knew what I didn't know. Now I know we are even.

> Now as I'm having some of this success, I'm finding that the income measurement is not as important to me. I'm

Your Hidden Credentials

beginning to think, "What really counts, you know? Is it important for me to get back up there, up that wealth ladder? Or is it more important for me just to survive and develop myself as a person?"

Elaine's, Joe's, Martha's, and Connie's turning points have several things in common. Each turning point generated a crisis that caused severe personal disruption. And each became a wrenching experience for them as they struggled to come to terms with the learning they had done and the people they had become. The experiences they had as they struggled through their turning points were difficult at best, characterized by strong feelings of dislocation, confusion, loneliness, and doubt.

But in each case, common elements also enabled them to deal with their turning points eventually with success. All four people gained in their self-confidence and their recognition of past personal learning. They recognized skills and behaviors they had learned, and they all learned new skills and behaviors. Ultimately, each of them transformed the turning point into a positive transition with a plan for further, conscious personal learning.

Turning points do not always have to be lost opportunities or wrenching experiences for you. Ideally, a turning point will stimulate a desire for more personal learning to augment things you have learned before. Far more desirable than missed opportunities or unsettling experiences is a consciously controlled chain reaction of personal learning and turning points. Turning points then would lead to a rich future of anticipated learning and change.

How you use a turning point assigns it a critical value in your development. It can be either an obstacle or a bridge to further personal learning. Ideally, a turning point will stimulate your desire for more personal learning. If you can learn from your turning points, you can begin to exploit them, changing them from missed opportunities or wrenching moments to successful learning experiences in themselves. Then your turning points can become the avenues for smooth transitions in your development.

Most people would prefer to take the jarring sets of circumstances, whether they come from within or without, and convert them into controlled turning points, or smooth transitions to the next stage of life. The next chapter, "Successful Transitions," illustrates that smooth transitions are indeed possible. The key to them is the ability to use the self-awareness you have gained to evaluate the impact of your personally learned skills. Then, you have a much better chance of recognizing and using your turning points productively as they occur.

3.

Successful
Transitions

Although personal learning affects you dramatically, often it is difficult to recognize because it accumulates one bit at a time, like snow falling. Recognizing your personal learning is difficult enough to do. But learning to evaluate it, harness its value, and use it consciously in your personal and professional development are the all-important next steps. When you can do that, many wrenching turning points such as those experienced by Elaine, Joe, Martha, and Connie can be converted into smoother, successful transitions.

Personal crises brought about the sudden turning points described in the last chapter. Like a slap in the face, these crises forced these individuals to face their respective turning points that awakened each person to their personal learning. Although their newly found awareness, their discovery of hidden skills and aspirations, eventually carried them through difficult times, Elaine, Joe, Martha, and Connie all would have preferred a smoother path to growth and change.

Of course, no one can eliminate unpredictable events or negative change completely—health problems, emotional conflict, and other such situations arise for everyone and they are unavoidable. But being aware of the shifts in your values, attitudes, skills, and knowledge can minimize much wrenching change in other areas of your life by promoting a more gradual evolution of skills and self-confidence. This is how you create successful transitions out of turning points. As you learn from your experience, you become better prepared to cope with the turning points that confront you.

The portraits in this chapter are shining examples of successful transitions. They describe the powerful personal impact that understanding your own personal learning can have upon you. Each person here has gained significant knowledge and has changed his or her values and attitudes through personal learning. Each of them has a remarkable story to tell. Although each story is different, there is a common element among them all. As the famous experiential educator John Dewey said, these people "have the precious gift of ability to learn from the experiences they have." As they make meaning from their personal learning, they become conscious and continual personal learners. Their learning is the key to the success that they enjoy.

Joan Miner

Joan Miner acquired her personal learning gradually, almost barely noticing it. She has become an extremely successful businesswoman in Des Moines, Iowa. But, it was not in her life plan to go into business; nor did she study business, finance, or personnel management in school. Her undergraduate education was as a secretary, interior decorator, wife, and mother for twenty years. Over time, though, she got the itch to do something more and, using her personal learning, she became much more than she ever dreamed. Her "graduate program" was on the job.

As Joan tells her story, she describes her personal transformation. After a career as mother, community volunteer, and paid worker, Joan reached her own turning point. She decided to ". . . be in business for myself." Joan drew on her past personal learning because she was compelled to learn more—quickly—in order to stay afloat.

"It's flow from that point . . . to today. I have taken hundreds of tiny baby steps . . . I changed a great

*deal, but the change is made up of those tiny baby steps
so I didn't notice it while it was happening . . . As a
human being with skills, it's the difference between
night and day."*

I was born in a very small town in Iowa, 600 people, in
1932. It was my universe. Women were housewives, teachers, nurses, and secretaries and that was the only exposure I
had.

In high school I was very much of an activist. I did everything they had. I played basketball, I was a cheerleader,
president of the class, president of the M club, editor of the
yearbook, and in all the class plays.

Then I went to Sioux City to college for a year though I had
no real idea of what I wanted to do. My father did not believe in higher education for women, but I had one high
school teacher who insisted upon my going to college, and
arranged for me to take the tests. I had a full tuition schol-
arship, but I still had room and board to pay. My father lent
me some money, and I worked as a secretary to the vice
president at 35 cents an hour. At the end of the first year I
was greatly in debt. I decided to go out and earn more than
35 cents an hour so I could get some of the debt paid off
and go back to school. Obviously, I didn't ever go back.

I found the working world a very exciting, rewarding place
to be. I started out as a secretary and did that for many,
many years. I was a legal secretary at a lot of different times
during my life. At that time I enjoyed the work tremendously. It was challenging. You were learning something,
working closely with an intelligent person, and I enjoyed it.

Then, the lawyer I was working with came to Des Moines to
the legislature and asked me to come down for the session.
Well, I liked Des Moines. It was bigger and more exciting
than Sioux City, and I decided to stay here.

One day, while I was walking across the street, I saw a
young man and literally fell in love on the spot. I didn't
know who he was or what he did. Two or three days later, I
went for an interview in a law office and, while I'm being
interviewed by the partner, in walks Bill. I realized he
worked in that office so I accepted the job for two hundred

dollars a month. It was unbelievable. I worked there for over a year. Of course, I started dating him, he was single, and eventually I married him.

After we were married, I continued to work for the firm—I hadn't gotten out of that rut. But Bill was making more money, so I started working part-time. Then, he began to make even more, so I retired, thinking, "Oh, I've really made it." Yet, I knew within three months that I hadn't and that I had to have some other kind of challenge. I very quickly got involved in special projects. I'd go back to the legislature and work for a session, or I'd set up a political office—short-term things, not a career by any means, but interesting, different things.

I got involved with Planned Parenthood and became very, very active in that. I went through all the chairs and went on the national board and executive committee. I learned a great deal about money in that context. As a housewife, I was handling thirty or forty thousand dollars a year at that time. The local organization had a budget that grew during my time from several hundred thousand to a million dollars. It really started making me understand a lot more about money because I felt a great deal of responsibility. On the national board we were dealing with ten million dollars and up, and it just kept expanding.

At that point in time I was really getting restless. My children were getting older and I had had just a taste of something more than taking dictation and typing. I had done some decision-making, and it had sort of spoiled me.

"I remember a particular night, maybe ten or twelve years ago. I'd had a dinner party and a lot of the people had gone home. Two men were still there. The three of us were philosophizing and one of them asked me, 'What is the biggest frustration or unhappiness of your life?' You know, nobody had ever asked me that before and I had never thought about it, and it just slipped out, 'I have no control over my life.' After I said it, I began to think about it. It was most definitely a turning point in my life when I began to take control."

Your Hidden Credentials

Joan's turning point at this time in her life did not jeopardize her livelihood or threaten the survival of her self-image like the wrenching turns of events that Elaine, Joe, Martha, and Connie had faced. Instead, she realized that she needed more for herself in life. Joan's practice in weathering earlier turning points by continuously seeking new learning put her into an excellent position to convert this one into a transition toward greater personal achievement.

About that time something finally got through the brain and I realized that I had made money for everybody I had ever worked for and that I enjoyed responsibility and making decisions. I just thought I ought to be in business for myself. I considered design, but I realized that there wasn't a big market in Des Moines for the kinds of things that I was interested in doing. So, I discarded that because I knew I didn't want to play. I wanted to take a serious business and make as much money as I was capable of making.

Then I really started to use my head and started to think, "What are my contacts in this city? What would be a place where I could get a good start?" My contacts were with lawyers, realtors, and bankers, so I just started looking for types of business that would relate to those people. The only one I could come up with was abstracting, but I kept discarding it because it didn't sound interesting to me.

Everytime a piece of property is sold in Iowa, somebody had to do an abstract continuation of the title. We are the only state that still survives on abstracts totally. Title insurance is illegal here. I had had no experience with the field. I had never worked in an abstracting office and I didn't understand how it was done, but I knew I had the contacts and that I could hire the people to do the work to make it go.

I had to understand motivation and I knew mine: to make it a success, to make money, to be a successful business person. The next thing I did was talk about it a lot. My husband Bill did the negotiating for me at the time because he had dealt with the abstracting companies as a lawyer. I did a lot of research by contacting successful people in other counties in the abstracting business. That way I learned what the problems and pitfalls were going to be. And so, I bought a company.

At that point in time, I had hired one person in my life and I had never had the responsibility of letting anyone go. I'd never had to fire anyone, but within two weeks I knew that I'd have to start letting people go. They simply did not do the quality of work that I needed to have done. So, I started reading everything I could get my hands on about management. I went on a self-help course and I started educating myself by reading, by going to seminars that I knew I needed, and by hiring people.

It's hard to interview people for a position you don't understand. I used a lot of gut reaction on that. Last year, when I started to analyze how long people had been with me, something like 65 percent had been here the whole five years. We've gone from 10 percent of the market to 40 percent, and from being a company that lawyers didn't respect to one that does the finest work in Iowa.

The first year we spent getting things going in the right direction. I thought it would take longer to change the reputation, but it didn't. The second year I tried to buy a company in an adjoining county, but they wanted too much money. So, I set up a whole plant over there.

In Iowa, you have a fifty year set of records from the courthouse. It was a very, very expensive proposition and I hired a computer company from Denver to do it on a computerized basis. I spent a lot of time making that decision because no other abstract companies were computerized. First, I went to Florida where the companies had done this, and found a company to look at that was our size. Then I went to IBM school so I could learn to talk computerese.

After making the decision on equipment, software, and what personnel we were going to train for what jobs, I started to look around for something else. I wanted to buy this newspaper, *The Daily Record,* but the gentleman who owned it was not interested in selling it at that time. So, I bided my time and we started doing all sorts of diversified services with the computer. We set up a closing company that does real estate closings, we developed and started selling a mortgage report, and we started a service for appraisers.

Then, I decided we had to make some progress on the newspaper. This particular paper dealt exactly with the same information that we did. The subscribers and people who paid

Your Hidden Credentials

for the legal notices were our customers. It was just a natural marriage. When they turned me down again, I said, "I think I should be honest with you. It is such a natural combination with my other business that I am going to be looking into the possibility of starting a competing newspaper." He immediately said, "In that case, I'm interested in selling." I made a proposal and in two weeks the paper was mine.

During her early stages in business, Joan came to grips with what she wanted out of her career—her own aspirations and expectations. Then, she embarked on a remarkable learning odyssey that blended actual experience, seminars, and reading. The areas in which Joan developed skill included corporate and personnel management, title abstracting, marketing, and computers. Through her personal learning, Joan preempted the negative aspects of her turning points and turned them into moments of growth and successful transition.

"One way I take stock of myself is when someone says something to me about how they perceive me and all of a sudden I think, 'Me. They see me that way.' And I compare that image to the way I see myself."

When I look back on it, it all seems so clear. I wanted to feel good about something I was accomplishing. I was doing this for this group and that for that group and nothing for me. That was my frustration. This work has most definitely been for me.

As a human being with skills, it's the difference between night and day. When I took this company over, I had never attended a staff meeting of any kind. I was at a national board meeting in New York and I woke up one night. I'm a very sound sleeper and it's unusual for me to have trouble sleeping. It was about 2 o'clock in the morning and I started thinking about what responsibility I had just taken on. I had just borrowed more money than I had ever thought about, and I wondered, "How am I going to swing this?" I turned on the light, got out my paper and pencil, and made a list of all the things that I knew I had to do immediately;

the concerns that I had; what kind of insurance coverage did the company have; and so on.

Then, I thought, "I bet all those employees"—there were 12 at the time—"I bet they're wondering about this. I suppose that I ought to call a staff meeting and tell the staff what is going on." That was my first one. I had not been exposed to anyone's management style and I had an awful lot to learn.

It's a flow from that point, ten years ago to today. If I really want to stop and analyze I can see just little baby steps towards where I am now. I have taken hundreds of tiny baby steps over the past ten years to get myself to a point where I could be independent. Now, I see myself enjoying the power males have had for so many years. It's been a great time to be a woman in business.

I've changed a great deal, but the change is made up of all those tiny baby steps, so I didn't notice it while it was happening. One way that I take stock of myself is when someone says something to me about how they perceive me, and all of a sudden I think, "Me. They see me that way." And I compare that image to the way I see myself.

Throughout her life, Joan exhibited an innate self-awareness and confidence in her ability to cope with whatever situations life presented her. Instinctively, Joan was consciously able to put her ambitions and goals into a deliberate plan of action. She identified her past accomplishments, including her skills with money, financial and legal people, and especially her self-acknowledgment of an ability to learn new things. She used them all to achieve what she wanted.

As she grew, Joan reached several turning points, including one that most of us feel sooner or later, the sense of a lack of control in her life. Her answer to this was an informal self-assessment and some decision-making about herself based on her newly recognized skills and desires.

Another turning point, the new responsibility of owning a company, caused her to wake in the night. However, her personal learning brought her through this internal anxiety by having her inventory both the problems she faced and some possible solutions to them—those things that she would do first.

Your Hidden Credentials

Joan succeeded well with her personal learning. She illustrates perfectly how to make smooth, successful transitions after many different turning points. She possessed self-awareness virtually from the outset of her life. She knew the skills she had learned on her own and she was able to use this knowledge to determine both future goals and the learning that would be necessary to achieve them. From Joan, we can learn how important it is to gauge new learning and to take stock of ourselves in an orderly fashion.

Bob DePrato, the subject of the next portrait, is also a successful personal learner. His learning began when he was in the army, giving Bob a strong sense of what would satisfy him, what would not, and why. He was too honest with himself to settle for anything less. As Bob sorted through a variety of careers, sampling and then discarding several, his personal learning kept him in the search for a meaningful career and a satisfying life. As he searched, his transitions were not as successful or as smooth as Joan's. But, he too eventually found a career that fit well with his behavior and his skills, firefighting.

Bob DePrato

As Joan Miner learned, she developed a clearer understanding and greater control over her life. She was strengthened and liberated. But Bob DePrato was stifled in his early adult years, a victim of his circumstances, powerless and controlled by the events of his life. For the first ten years after he left high school, Bob was a prisoner of his experience and expectations as surely as if he had been in jail. He believed that he brought nothing to the human equation, to life.

But when he came home after a tour of duty in the army, the travel, his reading, and his personal experiences had given him a ray of hope that ". . . somewhere along the line, something would work out." Bob struggled to align his work-life with the personal attitudes and vision he had begun to develop in the army. His personal learning kept him searching for a career that would also be an appropriate way of life. By his mid-thirties, he had developed a tremendous ability to analyze his learning after high school.

Bob is a short, peppery man. He radiates interest and energy as we talk. Bob still lives in Jersey City in the house where he was born. After his tour of duty in the army, and time in business and at the police department, Bob joined the Jersey City Fire Department. He is married and has three children.

"When you're programmed to be a loser for seventeen years, you don't stop being a loser just because someone says you're not . . . But they definitely left me with something, though I still had a confidence problem."

When I was 18, I joined the army. That was when I got plugged into the world because, before that, it was just Jersey City. I knew of no other environment. I didn't know any other cultures or customs or what people perceived as normal.

I went into the army just before Vietnam got hot. Kennedy had just signed a bill which gave married people a lower draft priority, so I had all these guys with college deferments coming in. As they graduated from college, they were getting picked off and they were coming because, although there was a war at the time, it was still a volunteer war.

I met an awful lot of people. I was a loser, a high school dropout and I wound up in a platoon with college graduates at a survey school. To this day I believe it was a screw up. I went to a school that required higher mathematics—you had to know trig functions—and I hadn't even taken high school algebra.

There were guys in there who had graduated from every college in the country, from San Jose State to Cornell, the cream of United States education. Two hundred men. Ninety-nine percent of them got drafted after their deferments were up. They were really sharp guys, but they were just poking through.

They were very good to me. I was 18 and they were 21 or 22, and I was struggling because the first thing the Army would say is, "If you flunk out of here, you're going to this field where it's 99 degrees and these guys are loading this 8-inch cannon (boom, boom, boom) . . ." I said, "Oh my God. It's like going to a Georgia chain gang." I was really concerned. These guys said, "Don't worry about it. We'll get you through it"; and they did. They just taught me what I

Your Hidden Credentials

had to know and I got through it. I just remembered the steps very mechanically and I got through the school.

I really got a lot from these guys. They took me around. They were a more elite crowd. At that time I didn't know anybody that played bridge or golf to begin with. They were what I considered to be exotic hobbies. These guys didn't slight me. They weren't crass. They were nice people. And they said, "You look like a pretty sharp kid. You ought to get . . ." I said, "What do you mean, a sharp kid? I dropped out of high school when I was seventeen years old." They all said, "You can change that."

They really gave me an indoctrination and some of it rubbed off because I liked them. They were nice. They handled themselves well and they were obviously all going to be successful. They had it together. The worst thing that happened to them was they got drafted. It interrupted their lives. Some of them were headed up the corporate ladder, others were skilled people. But they were happy with their lives. That was something that I certainly wasn't. I was just walking through it hoping something would come along. I had no choice. Either you get up in the morning and walk through it or you go home and blow your brains out. There's really not too many options. You either give up or play.

When you're programmed to be a loser for seventeen years, you don't stop being a loser just because someone says you're not a loser. It doesn't work that way. It doesn't happen that way. You think about it and you say, "Hey, maybe he's right. I hope he's right" but you just don't know what to do. We were together maybe ninety days or six months, then we were sent all over the world—to the four corners of the earth. They definitely left me with something, though I still had a confidence problem.

"The isolation drew me to reading . . . I had never read a book, an entire book, in eleven years. I never enjoyed reading. Over there, it was read or go crazy."

The isolation in the army is a big thing because it's like being in jail. In Korea it was just barren. I was in a compound with about eighty men. The isolation drew me to reading. It's like a typical story of prisoners going to jail barely able to read and for lack of nothing else to do, they read.

I had never read a book, an entire book, in eleven years. I never enjoyed reading. Over there, it was read or go crazy. I mean, when the money's gone, what do you do? I read anything, mostly what people might say is nonfiction because there was so much that I didn't know. And then people began plugging me into stuff. They would say, "Read this." And I'd say, "What's that?" And they would say, "Kafka. Everybody knows about Franz Kafka. If you want to know anything about literature, you have to read at least one of his books."

But I didn't get the prelude that the college student gets. When he gets introduced to Kafka, it's a formal thing through his professor. I didn't get that. I read it as a novel, period. I didn't look for the meaning. I read Kafka and maybe five years later I'm sitting in bed reading the *Times Book Review* on something that someone else wrote about Kafka and I start joining things together. I think, "Holy cow, this is wild." Then I go back and read another one for reinforcement or out of curiosity. This happened a lot. I would read something and get something out of it but not explicitly what the people who wrote it had in mind. It was like a disjointed piece of information.

Something else happened to me in the army. I picked up my high school equivalency. It was five hours out of the day and it plugged me into so many jobs. I often think about that. Five hours. You sit down and fill in a bunch of IBM holes and it turns your whole life around. I wouldn't be here today if I didn't have those five hours.

People don't achieve for that reason. They say, "I can't be this because I don't know how to do algebra," "I can't be that because I never took this formal learning," or "I don't have an affinity for that." That's become a big problem. They bought the mint. People cancel themselves out before they even inquire.

When I got out, I was thinking that I had to get a job. I never thought about a profession. I just thought about a good job because that's the way you're indoctrinated around here if you come from working class people. I had to get a lifetime job. Then, when I got married, that was it, locked in. You've got responsibilities, you've got to work.

I had a couple of jobs, decent jobs, but I said, "God, I'd hate to spend thirty years at this." First, I figured costs for a pipe fabricator. It was okay, but it didn't pay that great, so I went to the utility. That was a lifetime job because utilities don't go out of business. So I said, "Okay, I'm set for life."

Fortune smiled on Bob when he met those men in the army who helped him and convinced him of his ability to survive and learn. And, regardless of his attitude at the time, he had a sufficient sense of his future to complete his high school equivalency program.

But Bob's successful transitions really began with his appreciation of his own personal learning. While it was not immediately applicable to any of his employment goals, Bob's reading and the confidence he'd gained when the other soldiers helped him were instrumental in his long-term success. His personal learning gave him a foundation of self-confidence and expectations that wouldn't allow him to remain content with unsatisfying jobs, even if they promised him a comfortable living for the rest of his life.

When I came home from the service, I found out that my brother was a cop. We never liked these kind of kids that had cops in the family so I was shocked. He took it like I would take a job—someplace to go. He said it wasn't bad and it paid the rent. I thought maybe he had something there, so when he told me they were going to have a police test, I decided to try it. I did very well. I came in eighteenth out of a field of several hundred. I said "goodbye" to the utility because the increase was an immediate $2,500 a year plus the liberal vacation, clothing allowance, and all that. I just had to go.

I stayed there three years and the job became totally unbearable. It was almost becoming a social embarrassment for me to say to people that I was a policeman. I could never identify with the job. I felt like a centurion for the Roman Empire. I was never a cop advocate. I felt like, every time I went on call, the way I had to handle the call made me the advocate for the institution and never the person.

I worked in a poorer district and I'd go to a store and somebody had been beat out of their money. They would have signed something you'd need 20 power binoculars to read on the receipt and they couldn't read anyway. They just got beat. It was clearly a fraud, but not a fraud legally—just morally. The man takes people. He charges them 30 percent interest for inferior merchandise. And I'd say "No wonder they hate us, they're supposed to hate us. We're the enemy."

Those calls and the other calls were like a "no win" situation. You'd go to a house. Maybe a battered wife. The only threat to her life is attached to her husband. She wants him to stop beating her but she doesn't want him in jail. She just wants him to be nice. You can ask him to leave or you can lock him up on her complaint. Those are the only options. It's an impossible kind of a situation.

I had an uncle who was a fire chief and he's telling me, "What do you want that for?" I said, "I don't want it, but I can't leave. What am I going to do?" He told me there was a fire exam coming up and so I took it in 1970. Again, I did very well, so I took that job and I've been very happy with it. It's the greatest job in the world. It's like performing in the circus. You can't believe how great it is. You go out to a fire and, most of the time, even in the poorer sections, people appreciate you. They don't see you as the enemy.

If you take someone out of a window at a major fire, well the personal gratification alone, it's beyond words. Somebody is out there walking around because you were there. They're alive because you were born. That's what it amounts to. You just hope they'll do something good with their life.

Bob didn't credit himself with much self-worth during his youth, but over time he developed it to such a point that he could not ignore any potential opportunities that came his way. He possessed the confidence to leave comfortable situations and seek new jobs, and each new position taught him significant new skills. His attitude and wealth of personal learning made changing jobs easier and easier. Through these experiences, Bob learned to create his own constructive turning points to go along with the potentially wrenching ones that everyone encounters, because he saw opportunity through change. The more changes Bob created and experienced, the better prepared he was to make smoother transitions centered around future turning points. Eventually, Bob found a profession that suited him because he refused to stop looking, and he had overcome the fear of change.

Peg Moore's story offers a third example of how to achieve successful transitions. Unlike Joan, who chose her new career and, in so doing, constructed her successful transition or Bob who, armed with his earlier personal learning, searched for the right fit, Peg found herself starting a challenging job without any of the formal credentials required for the task. By accepting the job, Peg demonstrated confidence in her ability to learn what she needed to know in order to do the job well.

Peg Moore

Peg Moore's two children were asthmatic and she wanted better health care service for them. So, she attended a meeting on community health care issues in 1971. Within a year, with only a high school diploma and a commitment to the area where her roots ran deep, Boston's Old North End, Peg was the director of a new health center there.

"Although I was informally educated, I had enough background to get me through a lot of difficult periods. Plus my background as a bookkeeper helped me with the financial part—that was something I was good at. It was the health part that I was uncomfortable with . . . I managed to get through some very difficult times and pulled the center out of great financial difficulties. The programs were able to grow under my direction and so did the budget and the finances."

I did very well in high school. My father wanted me to go to college, but I had absolutely no desire to do that. I wasn't dying to get out of school because I couldn't stand it; it was just that none of my friends were going to college. I've lived in the North End all of my life and there is no pressure, no stimuli in the community to go to college. It's not really considered to be such a big deal.

Like all my other friends, I went to work and loved working. I met my husband and we got married about a year later. I got pregnant right away and had my first child two weeks before I turned twenty. My second child was born about a year later, and the third a few years after that.

I worked at a bank from when I was a junior to just before I got married. After that, I worked at a theatrical agency as a bookkeeper until my pregnancy would not allow me to work any longer outside of the house. Money was very tight at that time, so I had to find bookkeeping work at home so I could take care of my babies and bring in some money.

When I had my third child, I was still working at home. I became very interested in the health care of the North End community because two of my children were asthmatics. I found it extremely difficult to get help here for them. One day, I heard there was a meeting in the community around health care issues, so I went to it. I got very interested in trying to get better health care here, and began to go to those meetings regularly.

Eventually, the group asked me to chair those meetings. I knew absolutely nothing about chairing a meeting. I didn't know protocol and I still don't. But I did it. After working just over a year, we were able to open the doors of this health facility and begin providing services here in the North End. They were very much needed and very much used.

In 1972, the board asked me to take on the job as director of the health center. I was very uncomfortable with that because I was concerned that I wouldn't do the community just service because I had no education. But, I knew what the community would accept and what they would not. I also knew my limitations and I wasn't afraid to ask for help. Also, when I was chairperson of the board, we had a consultant doing a lot of the legwork. He kept me abreast of what was going on so that I was doing some of the actual work from day one also. I worked in every aspect of the center's development including construction.

So, reluctantly, I decided to take on the role with the help of some friends in the health care field. They said they would tell me to get out if I wasn't doing a just service to the community by holding this job without a degree . . . if that was a hindrance to the health center.

I've changed a lot over the years since I got involved. When I began—when I tell people this they laugh because they know me now—I was shy, very unsure of myself, and lacking any self-confidence. Yet I can't tell you why.

I think I changed around the time I took this job. I used to be very much in awe of people. If someone were a doctor or a lawyer or a teacher, it was like, "Oh my god, they're better than I am." There are a lot of people who are like me when it comes to dealing with professional people. Working in this job, I found that everyone is a human being. You put your pants on the same way my husband puts his on. I had to stop being in awe of people, and begin to feel as if I were on the same level so that I could be successful at what I was trying to do. As a result, my negotiating skills with these people became much stronger.

It's a matter of putting myself in a position where I feel as if I'm dealing with someone as an equal. Early on in the game, they were calling me "Peg," and I was calling them "Doctor." Then, I got to the point where I said, "Well, I love a first name basis so you call me Peg and I'll call you Peter." The doctors didn't like that—they wanted to be called "Doctor," but I'm sorry. Unless you're willing to call me "Mrs. Moore," then I'm not going to call you "Doctor." That put us on equal terms when we were talking. When we're in the examining room, that's another situation because you're the professional and I'm the client. But at the table, that's a different deal.

There was something else I learned that was really important. I used to sit at meetings, listening to someone speak and I would get into a panic because I didn't know what in God's name they were talking about. I would think, "What am I doing here? I'm out of my element, it's beyond me, it's over my head!" It affected me so much that I thought I was going to cry or pass out or both. I wasn't understanding a damn word that was being said and I would get so frustrated that I would almost be in tears.

I would look around and everybody was like engrossed in the conversation, nodding, the whole bit. Well finally something gave me the courage at one meeting to say, "Excuse me, I don't know if there's something wrong with me, but I don't understand what you just said." Ninety percent of the people in the room said, "You're right, I don't understand either." I realized that in most cases nobody knows what the hell the speaker's speaking about but they're all embarrassed to say so because they don't want to sound stupid.

I've had similar conversations with doctors who work for me. We'll be sitting down at a meeting with the staff and I'll ask a question. The doctor will come off with this long thing and finally I'll say, "Excuse me, but I didn't go to doctor's school. Would you please speak English?" That's their self-defense, their way to make the show.

Peg equated her personal learning with personal change. She described dramatic changes in her confidence, self-awareness, and skills since that first community meeting. Beginning with her decision to attend the first meeting, Peg had confronted significant turning points and moved smoothly through several transitions in her personal and professional development. Her personal learning enabled her to hold that job and become successful in it.

She had become more assertive with other people, adept in the rules and the group dynamics necessary to run a good meeting. She was less "in awe" of other people, including the professional doctors she worked with. Peg became a competent health care service manager on the job, extremely knowledgeable about the services her center offered as well as its financial condition. Perhaps most importantly, whether she was in a meeting, one on one, or with herself, Peg had learned how to ask the right questions to get the information she needed.

". . . I know that I learn something every day. I know that I've learned something when I can take it and apply it to something else. Then I've learned it."

Now I'm going to learn a whole lot about a new area—nursing home care. Some of it I'll be able to apply to my knowledge from here and there, personnel policies, dealing with people, third party reimbursements. But, there are other components of nursing home care where I have no knowledge so I'm looking forward to learning them.

The sad part is that the degree—that piece of paper—is very meaningful to other people. It was really frowned on when I would send a proposal into a government agency and they would ask what my background was. I graduated from high school. Period. I mean it was like, "What? You don't have a master's degree?" They were thinking, "What do you have, a Mickey Mouse operation going on there?"

Your Hidden Credentials

You just can't seem to get your foot inside the door without the degree. I know a man who works in an insurance company. He has all kinds of certificates from school programs and conferences and things that he's done in the insurance business. But when an opening occurred elsewhere in the building, he applied and they said, "Well, you don't have the degree." They would take someone with a degree in gardening because they had the piece of paper and bypass the person who has the experience in the area where they're looking. It happens all the time. It's very frustrating.

Joan Miner, Bob DePrato, and Peg Moore all identified their personal learning and used it to create their current situations. They still seem surprised, even astonished, at the things they have achieved and the impact that the learning has had on them. Each of them developed new attitudes and learned new skills and knowledge that they use continually. Although the learning appears to be unexceptional when taken piece by piece, its collective impact on each person's growth and development is significant. And all three of them used their personal learning to assure smooth transitions in their lives.

The same is true for you and your personal learning. Personal learning stimulates personal change, in your skills, your knowledge, your behavior, and your world view. If, as Allen Tough has found, most people learn continually, then recognizing your personal learning is the first step towards taking control over it and adding to it.

But, it is one thing to recognize your personal learning and quite another to be denied its value. How would you feel if you recognized your personal learning only to find that the rest of the world didn't care? Remember Peg Moore. Despite her success, one specter haunted her: the lack of a college degree.

As Peg recognized her personal learning, she used her knowledge to take turning points and transform them into occasions for improvement and more learning. But, the lack of a degree weighed heavily on her mind. In truth, this is a common problem for thousands of people.

Simply recognizing and using your personal learning is no longer enough. Getting credit for it in the outside world is the next step. Somehow you have to develop the credibility, either personally or through external recognition, to have your hidden credentials accepted as legitimate learning.

Just remember, you are not alone. Most people who have learned skills, who are aware of their knowledge, and who are ready to capitalize on future turning points have been stymied by institutional prejudices against their personal learning. They bump up against the world of formal credentials with their hidden credentials and, in most cases, they find themselves on the outside looking in.

4.

On The Outside Looking In

We have all seen the picture of two children with their noses pressed against a candy case. Boxes of candy and gum lie scant inches from them. But the sign says "five cents" and they only have a penny each. Or, we've seen the image of the lonely person at Christmas time, looking from the street into a brightly lit room filled with comfort and friendship where a family is gathered around the fire.

In such pictures, the thin pane of glass in the window separates two worlds, leaving the people on the outside looking in at the objects of their desire: so near yet so far away. This is similar to the situation in which many personal learners find themselves: separated from opportunity at work and in college by the lack of sufficient formal credentials, in spite of their self-awareness of skills they've learned informally.

Up to this point, we've explored the nature of personal learning and its relationship to change and development in your life. The preceding narratives have illustrated that most people learn useful skills and life-directing values and attitudes continuously. Furthermore, they suggest that if you can understand your personal learning, you will become more aware of your self-worth and succeed at transforming many critical turning points in your life into smooth transitions to further learning and growth.

Simply recognizing your personal learning, however, is usually not enough. Getting validation for it is the next step. Whether it comes from an employer or from a college, we all need validation of our hidden credentials. Unfortunately, for far too many people achieving success as a personal learner is not enough to earn credit in colleges or to win the promotions they deserve in the work place. In fact, most businesses prefer practical experience, but they also demand a college degree as a prerequisite for consideration for top jobs. Yet, most colleges traditionally deal with theory, and few of them give credit for practical experience.

Personal learners bring knowledge with them, but most colleges ignore it, insisting on their own curriculum as the only path to a degree. But personal learners who have been setting their own curriculum and learning at their own pace, who have the confidence that self-awareness brings and know the skills they possess, think that colleges are wrong. Yet, like the children in front of the candy store, they are left on the outside, knowing of their skills while looking in at the credentials and rewards that colleges have to offer. This describes the dilemma of many personal learners like Ray Miller.

Ray Miller

Ray is a police officer in a small city outside of St. Louis. A solid man with a gruff demeanor, the first impression he gives is that of a cop. He, too, is learning from his experience and about himself. The respect he has earned among his fellow officers is reflected in his position as head of the Policeman's Association. But, as he approaches forty, Ray is less sure that he wants to continue doing police work for the rest of his life.

"I don't know if organized higher education is entirely relevant to my situation. I would like to have a bachelor's degree or a master's degree in my hand. It would probably make it easier for me if I wanted to find another job somewhere. But, if you strip everything aside, we are just probably brighter than many of the

people who have degrees. It occurs to me that if some-
body would sit down and talk with me and listen to me,
they would realize that, and I wouldn't need to have
those degrees."

I have been a policeman for going on twelve years. Being a
policeman is not a profession I could explain to you so that
you would understand. It's not a profession that any person
would understand unless they actually do it for a certain
amount of time. I think it takes five years probably before
you really begin to take a look around you and see what this
business is really all about. I've gotten to the stage, now,
where there's nothing really that I can see out there that I
haven't seen before. There's not anything that's going to im-
press me.

So, over the years I've begun to lose my enthusiasm for the
job. See, I'm kind of a warrior . . . a tired one. Tired of
being accused all the time, even by the people who are close
to me, of being an uncaring, uncompassionate person. I say
that this is not true. This job does it to you. You just have to
put that front out and keep your defenses up. I have seen
everything there is to see and I can't overreact. But I am a
caring person. I am a compassionate person.

As I grow older, I am more aware of these things. When
you get to be past 35, I think you develop more of an
awareness of yourself, of where your life is going, how it is
going to end. You get more careful and you become more
aware of the people around you and more aware of your
lifestyle itself.

"80% of the things I know about, I learned on the
streets . . . I know a lot, but I don't have the degree."

Sometime in the past, I became aware of how much I really
didn't know about things, so I began to read in an effort to
know. It made me feel more confident and it made me feel
that I could represent the Policeman's Association better. I
can speak my mind now. I am not slow-minded, so I can do

battle, but I just did not feel confident about myself. My mind is a collage of information that, you know, I can dig out and use one of these days. It has made me feel more confident dealing with people. And, that confidence itself influences other people.

My circle of friends includes a guy with a master's degree in history. He is one of the finest experts on the Civil War that I have ever met and he and I share a little bit of knowledge about sea battles and so forth. He is very bright. People wouldn't imagine a policeman with an IQ of 150. He is hiding and he knows it. Maybe I am too. But I feel . . . it makes me feel good that I can communicate with these people.

Eighty percent of the things I know about I learned on the streets. Those things don't fit directly into an institution of higher learning. But, the biggest thing that a college could do for me is, when I sent my resume to somebody, I could write on there that I have these degrees which I don't have now and they would call me in for an interview. I'm not the only one who feels that way, I'm really not. Is it going to be worth it to me to spend a lot of money on a traditional college education? Is it going to make me a better policeman? I'm talking about me as an individual.

I know some well-educated people and I used to think, "I'm not hitting it off with these people. We are not connecting somewhere because that guy is probably much better educated than I am and he is speaking in lines that I can't get to." Well, I have come to find out that it may not be true. Maybe I was speaking in lines that he couldn't get to, but it's not supposed to be that way because he is that much better educated than I am. What it all essentially means is that I just don't trust the value of an education.

I don't know if I am who I want to become yet. I don't know the answer to that. But, here is what I do know. I know that I am brighter than the average person around. I know that if I can take that intelligence and integrate it into my reading and then take the insight into people that I have gained over all of these years that nobody, hardly anybody, has ever seen. . . . I probably know more about people than I do about anything else.

Honestly, if you could combine those things with all the knowledge I have about how the world really is and some other things I might have going for myself, and my intelligence, then I've got pretty much what I need. I know a lot, but I don't have the degree.

One of these days, an opportunity will come along and I'll take it. I figure I have some things to offer and I have something that's real important, and that's integrity. Let me tell you I've been doing this for 12 years and I haven't taken a nickel yet.

Plainly, Ray Miller personifies a person who has learned a great number of skills, who is aware of his learning and ready to take advantage of opportunities to change for the better. His biggest obstacle is the societal belief that learning happens in schools and that credentials are validated only by credits and degrees. Ray's problem in realizing the potential of his hidden credentials lies in the inflexibility and remoteness of those institutions designed to offer him the opportunities of which he spoke.

Most businesses do not orient their recruiting programs to seriously include personally-learned skills, such as Ray's, without some college or university's stamp of approval, the degree. But, to ask Ray to return to school at the age of 35 seems to be as highly impractical as it is an affront to him and his useful skills.

If he decided to attend night school, the prevailing alternative for most working adults, Ray could easily be 45 before he received his degree. Although one of our goals is to learn consciously for the rest of our lives, the prospect of attending classes for nearly ten years while continuing a full work schedule is a major obstacle to most people. Furthermore, the program he would take probably would not award Ray any credit for the personal learning he had already done. In other words, he would start at the beginning, like an 18-year-old. Nonetheless, the bottom line is that if Ray doesn't submit to the limited menu that most colleges offer, the opportunity for which he is on alert might well never arrive.

The Adult Student: College Fear and Loathing
We met Bob DePrato in the last chapter. After years of struggling, first in high school, then in the Army in Korea, and finally as a member of the Jersey City police force, he found satisfaction working as a fireman. He liked the job and it gave him a real sense of security. Clearly, Bob has done an exceptional amount of

learning on his own and has been able to take advantage of many opportunities in his life as a result.

But, after several years on the force, including a promotion to lieutenant, Bob came up against the same wall that Ray Miller experienced. Without a college degree, he was at a dead-end in his job no matter how good he was at it. Bob could not advance any further.

"What happens is that at some point someone identifies that you have untapped skills and no one knows a better way to develop them than school. You never heard anybody grab anybody and talk to him like a Dutch uncle. Put their arm around him and say, 'Listen kid, you've got what it takes, you ought to take a correspondence course . . .' What they do say is that college will get you ahead."

We had a new chief of fire prevention. Most Jersey City firemen are local. His background was similar to mine. He was a golden-glover and he came on the department and went right up the ladder and became a Deputy Chief in a very short time. In between he dabbled in a college education and now he's right at the top. He's like running a corporation, he's got his degree, he's working on a master's, he teaches at the National Fire Academy, he travels all over the country as an advisor, and he's very successful and a nice guy. He gets to do it all. So, he convinces me that I've got to go to school.

What happens is that at some point someone identifies that you've got untapped skills and no one knows a better way to develop them than school. You never heard anybody grab anybody and talk to him like a Dutch uncle. Put their arm around him and say, "Listen kid, you've got what it takes, you ought to take a correspondence course." Nobody says you should go to the library and get a book—people just don't say it.

What they do say is that college will get you ahead. Like my boss told me, "You need this. It's the only way out. You're

Your Hidden Credentials

going to be wasted if you don't get this. You have to have it." He's telling me that they're not going to let me in without it. I don't care how bright you are, you don't get into the club unless you wear the ring. Out of all the all-stars, about 99 percent are going to wear the ring. That's where they're going to come from.

So, I finally dragged myself over to the Community College and took two courses. And, I was scared to death. I had been at war, in police riots, and a firefighter and that was all right. But, now I'm going to this college and I'm scared to death. I literally had stomach cramps—got physically sick. I really don't know why I was scared. I had no idea . . . no reason . . . it had nothing to do with anything . . . there was no consequence to pay. There was no overt, physical, explicit consequence. I could have gone there, flunked out, and nothing would have happened. I still would have been a lieutenant in the fire department. Nothing would have changed. As it turned out, I left there with a 4.0 after taking four courses. But, when I first went through the door, I was dying.

"I was sold this bill of goods about college, that everyone who graduated from college was far superior to the average guy walking the street. I don't say it's a conspiracy, but I think at some level people actually peddle that attitude to keep some people out of colleges and other people in them. Where I come from it's meant to keep them out because there's only so much room at the top. It's that simple."

Then, I came over here to St. Peter's, in a program that was especially designed for adults. You don't get that bureaucratic run-around at the registrar's office. They bring you in slowly, and by the time you get a four-year degree from these people, you've seen it all. You've seen how they do business, and later on they kind of let you spread your wings a little bit. They'll tell you to register for your own courses and what you need so you can get a taste of it. Then you know how they do business. They have a course, a workshop where they try to give you credit for your life experience learning. They ask you tó describe and document the things that you have learned from the things that you have done. I love it. Now, there's no stopping me.

I was sold this bill of goods about college, that everyone who graduated from college was far superior to the average guy walking the street. I don't say it's a conspiracy, but I think at some level people actually peddle that attitude to keep some people out of colleges and some other people in them. Where I come from it's meant to keep them out because there's only so much room at the top. It's that simple.

Unlike Ray Miller, Bob decided to continue his schooling. His initial experience, however, terrified him. Bob—a war veteran, a former police officer, a fireman, and an accomplished personal learner—was utterly intimidated by an institution that theoretically had been designed to help him. The very fact that he eventually performed so well in school shows that he had nothing to fear. Indeed, his ability to excel inside the institution indicates that he probably achieved a high quality of learning outside the institution also. Still, Bob finally had to go to another college to receive any semblance of credit for his personal learning.

Through Bob and Ray's experiences, we have learned about some serious problems with our colleges and other educational institutions. Far from encouraging the romantic ideal of the self-educated person, most colleges stick to a fairly narrow set of parameters in terms of what is acceptable for credit and study. In addition, as Bob's traumatic reaction indicates, many adults see colleges as awesome structures, restricted to only a few chosen people.

Since more and more people attend college every year, this latter perception would appear to be unfounded or unreasonable—but it is not to adult learners. Despite their own acknowledgment of their personal learning, many adults who have succeeded without college undervalue their accomplishments. Why? The answer can be found by examining the role colleges have traditionally played in our society.

The Role of College in America
During the last forty years, the world has changed dramatically right before our eyes. It continues to change today. These changes

include the structure of our population, our work force and the work that we do in an emerging world-wide economy.

One of the most noticeable and important of these changes regarding your hidden credentials is the shift in the mean age of the country. The American population is aging. Historically, we looked like a pyramid, with many more young people at the bottom and progressively fewer senior citizens at the top. With fewer new births, however, our population structure now looks more like a square, with more older people in proportion to the youth population.

In the past, employers and colleges could pick from an expanding pool of young people when they were looking for new skills or new students. Consequently, renewing your skills or increasing your educational levels was a low priority. If you were an adult, you were likely to hold the same job throughout your life. Continuing education was considered a nice thing to do if you had the time and the money.

But, with the "squaring" of the American population and a changing international economic structure, the demands on our work force are changing. We will increasingly be the inventors and servicers of ideas, and our jobs will be knowledge-intensive and technical. Change, and understanding changing information, have become central elements in our lives while only brief years ago, sameness was the rule.

If you undergo the average work experience of today, you will hold more than six jobs requiring different skills and capacities during your lifetime. More jobs in more places with different people requiring different skills, that is our current situation. As a result, your personally learned skills and attitudes and your ability to learn throughout life have become extremely important to society and valuable to you. You need to use them to understand and incorporate the diverse experiences of your life in an information-rich society. Educators and employers should use them as the keys to further learning and growth in your work as well as in your civic and personal life.

Our colleges have been buffeted by these winds of change. As recently as 1945, college in America was considered to be predominantly an academic and scholarly experience. The degree was a credential that verified achievement in that world. Since then, colleges have become an acknowledged part of the opportunity structure. Access and achievement, not birthright and brilliance, have become the dominant characteristics of our system.

The rapid growth of public and community colleges and need-based scholarships have been the hallmarks of the change. Colleges have expanded their services and innovated with their curricula in order to serve new students. Their purposes have expanded from a strict focus on academic preparation to a wider focus that includes economic preparation as well. The degree has become an economic credential as well as an academic one. Gradually, the marketplace has come to rely on the degree as the main credential for work.

But, there is one major problem: our colleges are still structured for young adults. For most young people, college was and is a "pre-work" activity. They need a place to live, to study, and to socialize on a full-time basis. They learn by day and study by night. Summer is the time to work and play. School is the organizing influence in their lives.

Most college curricula have been organized for these students, whose major prior experience has been school and book learning. Its purpose is to expose them to ideas, concepts, and information for the first time. The faculty member is the giver of information and the campus is where the learning occurs. Despite our aging population and colleges' expanded purposes as institutions of economic opportunity, their basic structures have stayed intact. They are daytime, full-time, instructional, and they are structured to prepare people to begin adult life. Consequently, they are not for adults, nor are they adult-friendly.

The Power of Personal Learning to Change Society
An extension of your personal learning beyond both skills and self-worth is the realization that many of the problems associated with college have to do with the institution, not you. The personal learners who follow discovered that while they could succeed in college, they could not give college high marks for fulfilling their needs.

When we first met Martha Lucenti, she'd described herself as a shell-shocked former housewife who had gone through a rather disastrous divorce. But, as she looked back over her experience as a wife and mother, there was determination in her voice. As a woman without a formal education or credentials, she showed confidence as she discussed her decision to continue learning rather than surrender herself to depression. Much to her surprise, however, her college experience taught her more about how society works than she ever expected from any classroom.

"I certainly did a lot of child psychology raising four kids. . . . Maybe I didn't know what the great philosophers thought, but I certainly knew that you handled one child in one manner and another in an entirely different manner."

Learning is a matter of survival. And, it's important to know the worth of what you've learned. I've learned that I had done an awful lot in my life that I just didn't really think amounted to anything or served any particular purpose. It had never been put to me that the things I had done were of value. It was just taken for granted that you did them.

Here I was, uneducated with only a high school degree, and I certainly did a lot of child psychology raising four kids. I was actually doing the child psychology without reading up on it. I wasn't knowledgeable as far as having a degree in it. But, through experience and hit and miss, I found out which things worked and which things didn't. Maybe I didn't know what the great philosophers thought, but I certainly knew that you handled one child in one manner and another in an entirely different manner.

You begin to ask yourself after a while, "What did I learn? What am I worth?" My literary works were *Parents Magazine* and that type of thing. I was not into *Romeo and Juliet* because I didn't have the time to read, didn't have the inclination, and once I read it, what good did it do me?

Romeo and Juliet is a different kind of learning with a different set of goals and values. If I talked to somebody that was into this heavy literature thing, I had nothing in common. People like me take a lot of what they already know, the knowledge they already have, for granted. They think it is nothing and they are wrong. Those who did not have a college education, we did it the hard way, learning by the mistakes we made and by the mistakes of our friends or parents.

Now, these same mistakes that we learned by doing are the examples they use in the books that they teach us from. When you look at it that way, it's even funnier. It's all basically the same knowledge, it's just there's a formal way of learning it and an informal way of learning it. I think it's better to have the experience first. You can understand it and cope with it and kind of relive it when you go to school. You say, "By golly, that's right! It really does work that way." Yet, if you haven't had the experience, you can only assume it works that way because you don't have first-hand knowledge.

When I finally thought about what I had learned, things that I thought were insignificant all my life became important. I had seen them just as a matter of life, yet they were knowledge. Seeing that knowledge is a great tool for learning that you don't pay much attention to.

All of a sudden, I was as good as or equal to those who had formal education. I took the experiences that I'd had and restructured them to a point where I was able to bring out the knowledge that I actually did have and put it in the same form as the others who learned from the books. But, I had actually gotten mine in real knowledge and real experience.

Still, even for Martha, the power of the degree lingered on. As she reasoned it, the degree was the lever, pushed by knowledge, that opened the doors to the work place. To her, it became far more than an issue of confidence and self-esteem. Knowledge without formal recognition became a matter of economic power. Recognition of her personal learning, her hidden credentials, became her goal.

"To tell them that I sat in the library and read all these books on nuclear fusion . . . is not going to do a thing. But if I have a degree . . . that will mean something."

The question is "What do you want to do with the learning?" You wouldn't want to go get a college degree just to say that you've got one. You don't need a degree to get more knowledge. You can go to any library and pick up a zillion books and learn that way, if that's what you want to do. There should be an ulterior motive or reason for getting one other than its actual worth, other than just learning more knowledge.

The degree is something that is of value to you. It is a tangible thing that you can actually show people and say, "Look, I have accomplished this much and now I can go further from here."

As I have been going through the working world, I've found it very frustrating to have the knowledge to do a job and yet, because I did not have the degree, I was not able to get that job as far as having the title or getting the pay. You could get a person who would be fresh out of college that had the degree and they would get the job having no experience. I would end up having to train them to do their job. This was an extremely frustrating spot to be in: I was not a degree holder so I did not qualify.

It hurt. When I was in high school, if you got your diploma that was enough. Then, it got to the point where if you didn't have your bachelor's, there was a problem. Now, it seems like everybody is talking about master's degrees. I think a lot of people today that do not have degrees but who do have common sense and working experience have found out that wasn't what the employer was looking for. Employers put a high premium on the degree these days. I myself would not only like to get an education but I would also like to get ahead. And, you have to have the formal degree if you are going to get anyplace.

To tell them that I sat in the library and read all these books on nuclear fusion or American history or whatever is not going to do a thing. But if I have a degree which says that I have done this and I have gotten a grade from the professor, that will mean something.

Yet even as she recognized the hard reality of the power of the degree, Martha insisted on the importance of recognizing personal learning and the people behind the learning.

You could have a degree and not know anything and probably you would go a little ways, but you wouldn't go very far.

On the other hand, you could have a lot of knowledge and no degree and the same thing would happen. But, if you have the knowledge and the degree then you could probably make some tracks. It would be the same thing as getting in the car and just going around the same block. You are not going anyplace. Or, you can get in a car and have a little bit of knowledge and go where you want to.

Someplace in the world they have put such a high price on the degree that anything less is not valuable. I am saying that is wrong. We have to have some of everybody. You have to learn to take people as human beings rather than slot them into areas. There're good bricklayers and bad lawyers, you know, so just because you are a lawyer doesn't make you something that is extra special and wonderful. That goes for just about any profession or field. I think it's a shame to think that somebody who doesn't have the degree feels less than the person who already has it.

I know some people who have them who I don't really care for because they feel that once they have it, they are above everybody else. They lose perspective of who they really are and what they really want. I think it all boils down to no matter who you are, or what education you have, you are a human being and you have to treat others as human beings. Life is all channels, not just one channel, and you have to learn to deal with it.

Martha understands that standards of learning are necessary as measures of the quality of your education whether you get it from a college or on your own. But she also realizes that college as it is organized today is not designed for the adult learner.

Roadblocks for Adults in College

Too often, college is a series of frustrations for the returning adult learner. Some of them are petty, but very real. Others are more damaging.

- The bookstore closes at 4:30 when you can't get there until 5:00.
- Your academic advisor doesn't have evening hours and you work during the day.
- The library has only limited hours on the weekends when you can study.

- Continuing education courses are offered at night, but they are not part of the regular degree program, so their credit does not count.
- Many of the courses that you need are scheduled during times when you can't attend.

The combined message is clear: this college was not designed for you. Other policies are demeaning and downright damaging to you as a learner.

- Regardless of what you already know, you find yourself sitting between two recent high school graduates in front of a teacher who is talking about something that you do every day at work.
- Smart high school graduates get advanced placement if they can score well on tests, but there is no way for you to get advanced standing in your areas of work experience and personal learning.
- There is enough financial aid for qualified full-time students, but it seems much harder to get what you need if you are part-time.
- There are late fees and other penalties if you enroll late.

But, the loss of the learning that adult learners already have to offer colleges is by far the worst consequence of all of these restrictive policies to adult learners. Not only has society lost the resources of adults who have learned outside of college, but the colleges themselves have lost the opportunity to improve their institutions through the experiences of adults. Again, this loss is underscored most eloquently by an adult learner who nearly was disenfranchised by the biases of college and business.

Peg Moore, extraordinarily successful as the founding administrator of her health care center in Boston's North End, admitted that her lack of a degree raised questions in the minds of professional and funding agencies during the early part of her tenure. The questions always seemed to connect her lack of formal preparation with the credibility of the program, even though all available evidence pointed to success.

Eventually, Peg decided to get her degree. Her decision was a classic blend of practicality with finding a college that respected her experience and personal learning. Peg needed ". . . the piece of paper." But, she acted upon her decision to go back to school only when she found a college that was friendly to her regarding her situation.

"The actual deciding factor in going was that I learned about a program that allowed me to use some of the things I've already accomplished toward the degree."

I don't know what the future is going to bring and I saw this as an opportunity to be able to do something about it for myself. I found it to be a tremendously exciting experience—I liked it very much. It was very difficult to hold down a full-time job, go to school (which was considered to be a full-time program), take care of the family, and do it all in fifteen months. I literally dragged myself to school, and I just about cried in the classroom to get through it.

I'm not sure what, in the end, prompted me to go back for a degree. I look at myself and know that I am somewhat of a hypocrite because I wanted to go back and get it, even though I wasn't sure what the degree was going to teach me. Not that I, in any way, think that I know it all and can't learn—that's absolutely not true. I learn every day of my life. But, what were they going to teach me that I could benefit from on this job?

I see so many people coming out of school with a degree in their hand and, quite frankly, they don't know what to do with it. They've learned in the books, but it's not the real world. The book learning is only a base and kids coming out of college don't know what to do with it.

Let's take determination of need, for example. The book says you fill out an application like this and you file it over there and then you take it here and there—through the process—that's what you learn in school. The real world is that the piece of paper is only one small part of it. Like, that piece of paper is a waste of time because nobody ever reads it. They turn around and ask you ten times the same 80 questions, each of them ten times. You have to depend on the personalities of the people you go before in the process. That's the real world . . . the roadblocks, the setbacks.

Your Hidden Credentials

Then, there was my future to worry about. In other words, if for some reason I decided to leave the health center, maybe I would not even be talked to by other groups because I didn't have the piece of paper. Even though I could prove I had done it—it doesn't make any difference without the piece of paper.

The actual deciding factor in going was that I learned about a program that allowed me to use some of the things that I've already accomplished toward the degree, so it wouldn't take that long. Therefore, I wouldn't have to go the formal route of a four-year program with four courses a semester to achieve the degree. I didn't have the time, the patience, or the inclination to do that.

I was able to accomplish a lot without the paper, more power to me. Some of it was luck, pure luck. But, I think the important component of this whole thing is that if you have an education with no common sense (and there's nothing in any school that I know that teaches you that) then you really do have a worthless piece of paper. I think some of it has to come from either personal experience—the real world experience, in some business venture, be it when you were younger or whatever, and then apply that to an education, then you've got a good combination.

Colleges need to change also. I'd want courses that were going to be related to real world situations. They need to be taught by people who have done it; not by educators. Also, I think that schools make a big mistake when they teach things that they don't apply to themselves. How can you have any respect for an education when the people educating you don't have the sense of management control, organization, or even a sense of decency in how they treat students?

I'm sitting in the classroom and I'm learning from you how to do finances. Okay, because you're the expert at it, I'm learning from you. But, my friend . . . , I could be standing in that classroom and you'd be sitting down and I'd be teaching you organization. You have expertise that I don't but you have to respect that I have expertise that you don't. If you know that, you will have respect for my time and, quite frankly, both of us can walk out of there having learned something.

When I'm on my own, I learn from people. I'm not a particularly good reader. When I go to meetings, it's at the breaks and after the meeting that I get the wealth of information, because I talk to you and you tell me what you're doing. I've picked up an incredible amount from people that way. I know I've learned something when I can take it and apply it to something else. Then, I've learned it.

We're taking a lot of people in here without formal education and teaching them how to do certain things. They've become so good at it that I'll match them against anyone. Talk about mental health counselors or mental health therapists. Take the nondegree people and match them against the degree people, and I would take the nondegree person much of the time, certainly, in straight skills of how you work with mentally ill people.

It can be learned. But, the point is, whether you learn on the job or on the street or in the classroom, if you have the talent and the common sense, you can do it. You can apply it. You can make it work. There is no school that teaches you common sense. Take my case. I knew the community, and I cared about what happened to the community and the people in it. You can't learn that part very easily. It takes a long time, and you've really got to want to learn it. The other part, the health care part, the academic part, can be learned quicker and faster. I've seen it in other areas of this community and in this building with other jobs.

I can't imagine going to a marriage counselor who isn't married and has no children who says "Are you having problems with your children?" and then tries to give you advice on how to deal with them. They haven't gone through it. So, in most cases their advice is going to be limited. I'm sure a few could do it. But, how could you possibly imagine what it's like to live with a teenager?

Before they opted to attend college, both Peg and Martha couldn't validate the hidden credentials that they had earned over the years away from school. Each woman knew that she possessed new capabilities as a result of her personal learning and life experience, that she had become a different person. Yet, when they did go to college, both of them were exasperated with an educational system organized to provide instruction and degrees to recent high-school graduates.

As they struggled with the system's denial of their personal learning, they were angered by the social as well as the economic impact of that denial. They possessed something of value and they couldn't get it appraised. They couldn't determine its value, and they couldn't use it as a base for the additional learning and work that they both wanted to do.

Both Peg and Martha had developed tremendous personal assets that had no formal value in the world they were entering. They were millionaires, but they had the wrong currency and there was no rate of exchange.

Clearly, we have changed, as has our society, our economy, and our expectations. But most of our colleges have not changed. With their old structures and policies, they are caught between new clientele and changing societal needs. They have become gatekeepers to the marketplace, regulating instead of ensuring your access to it.

An adult-friendly college℠ knows that, for you, an education is only one part of a complicated and busy life that includes personal commitments, work, and family responsibilities as well as school. It is accessible to you personally as well as being responsive to your needs and your life situation.

Some colleges have chosen to become more adult-friendly. They are establishing programs and services that are precisely designed to be more responsive to the ever-burgeoning population of adult learners who have learned significantly on their own. In the next chapter, we'll see exactly how these colleges do it.

5.

Adult-Friendly Colleges™

P_{eg} Moore and Martha Lucenti have given us some telling insights on the contemporary college experience for adults returning to school. Martha wanted official recognition of the talents and skills that she brought with her to school. And she stated emphatically that college was the only socially sanctioned institution that could confer the universally accepted document, the degree. Although she recognized that there had to be standards of quality for the recognition and gauging of personal learning for credit, Martha emphasized the importance of acknowledging the acquisition of useful skills outside of the formal educational setting.

Peg Moore identified another area of importance to adult learners left unaddressed by most colleges, the exchange of knowledge between teacher and student. Peg was proud both of her ability to learn new material and her awareness that she brought a substantial body of knowledge with her as a student to the classroom. There were things that she could teach if the atmosphere were more favorable. This balance of experience and knowledge between teachers and their adult students demands a different kind of attitude in the classroom. In one sense, Peg has intuitively created a classic educational ideal in which the roles of teacher and student can be interchanged.

Millions of people like Martha and Peg need the same kind of visible credentials to go along with what they've learned on their own. They need them in order to be franchised for seeking a better livelihood, but they also need a learning experience that will meet their special requirements as adult-learners. The profiles that follow illustrate many of the attitudes and requirements that colleges must have if they are going to be truly "adult-friendly" in the future.

Betty Hill

Betty Hill shares a background similar to many of the persons you have already met in these pages. Instead of ever considering college seriously, she married shortly after high school to raise a family. Through the years, Betty gained some experience as a file clerk and typist, as a farmer, and through organizing her community's library. She never had given much thought to preparing herself for sudden turning points in her life, nor had she considered her own ability to learn until her husband suffered an injury that left him bedridden for an extended period of time.

Then, Betty began to think seriously about going back to work. But, she was confronted by serious obstacles to her plan. Betty needed to find a college that respected her personal learning as well as her complex scheduling needs as a mother and community volunteer worker.

"Then there's the time, the money, the logistics of school, traveling back and forth, the hours—all those types of things that become concrete facts when you are an adult trying to go back to school. You've got to deal with them. They're not pie in the sky."

We were kind of coasting along without too many waves disturbing the surface, when my husband had a bad accident that put him in the hospital for almost seven weeks. He broke his hip falling off a horse.

We had always assumed that at some point I would go back to work. But we didn't know quite when and I didn't know quite what. It was kind of an unspoken agreement that when I went back to work, I would be doing something that I enjoyed, but probably not full-time because I had the farm and the children and everything. It would just be for a little money and to get me out of the house. Well, we started talking when he got home from the hospital and I said, "I've got to start preparing for something."

I didn't know what I wanted to do, but I knew that I didn't want to work in the library system because the openings were few and far between and the pay is low. And, I didn't want to go back into the business world. That was fine and it filled the gap when I needed it, but it wasn't something I needed to do any more. I was tired of being paid to do someone else's thinking.

I felt that, when I went back to work, I wanted to be paid and acknowledged for my thinking and my results; not for someone else to mouth what I had researched and typed up. So, I went back to the Community College of Vermont and took one course in Creative Writing. While I was there, I began to do a lot of thinking about the future. I decided that I liked working with people and I liked the idea of teaching and working with adults. Teaching children would be for someone who had more patience than I did.

I liked the idea of what I was seeing at the college; people coming in. All levels of people. In that class, we had everyone, from a teacher taking a sabbatical to a woman who had just barely gotten her G.E.D. and who said, "I'm not stopping here." I thought it was great.

We had people there who had not done too well in high school because it was just something to be gotten through. They were finding themselves coming back because suddenly they wanted to know more. They wanted to learn more. They wanted to be able to express themselves. So, I thought, "This is terrific. This is really something that, even if you are a little down, you can walk into that room and you know that they are counting on you. They are waiting for you to help them pull it together."

"Flexibility is the key. Colleges need to realize that they are not dealing with an eighteen-year-old who has to be told, "You have to do this." Like threatened or

cajoled. You are dealing with people who are sometimes
older than you and sometimes wiser than you too, in a
lot of respects. . . . There has to be a mutual respect."

There was a lot of openness at the college. There's lots of different kinds of ways to study and learn, lots of programs. You can work in more than one program and you don't get locked into one way of thinking. Some schools try and push different philosophies and you are expected, if you're working there, to conform to within a certain degree of their thinking. With adults you need to be much more flexible.

At the Community College of Vermont, I spoke with a counselor and she said, "Why don't you try the Educational Assessment Course over the summer? You have a lot of things in your background that will probably be worth credit and be valuable to you." That meant a lot to me because I didn't have a great deal of time on my hands and I had other things that I had to do every day besides just go to school. Then there's the time, the money, the logistics of school, traveling back and forth, the hours—all those types of things become concrete facts when you're an adult trying to go back to school. You've got to deal with them. They're not pie in the sky.

Flexibility is the key. Colleges need to realize that they are not dealing with an eighteen-year-old who has to be told, "You have to do this." Like threatened or cajoled. You are dealing with people who are sometimes older than you and sometimes wiser than you too, in a lot of respects.

There has to be a mutual respect which I don't think the teachers who deal with the younger group understand when they get the older group. They still feel this need to motivate and they may be a little lost and overwhelmed when they meet a group that is already motivated and is just waiting to act and respond and absorb.

I think you have to train teachers specially to work with these adults. We aren't going to be put in lecture halls and told, "You are number 76, you are number 115. There are 300 seats up there and they are all going to be filled up on Tuesday night and Mr. So and So is going to stand up there and give a lecture for forty-five minutes and you will absorb everything he tells you."

Adults are generally not going to do that because they've been out of school too many years and they have some experience and knowledge. They need the give and take of questions and answers. They need smaller groups where they can feel there is no crime in asking a question; large groups intimidate people from asking questions. It's strange how one question will spark twenty or thirty others; but try to get that first one asked, it's hard sometimes. Everyone feels, "I'd like to know but do I dare ask?" In a smaller group, that hesitancy dissolves faster.

I've got a lot of strikes against me. Probably a lot of these jobs that I go in to talk to people about will be occurring when I am forty and beyond. I'm 36 right now. I have to have an education behind me so that no one can question me when I come in and say, "I can do this."

There is a gap between structured learning and life. We need other ways to show learning besides a degree, other ways to demonstrate that you are competent to do a job. For instance, people aren't given acknowledgment for the fact that they attend maybe twenty, thirty workshops over a period of time that qualifies them for much more than standing doing one simple task. And, when there is no acknowledgment, there is no self-esteem.

That is why a lot of people feel that if something happens to a particular job situation, they do not have alternatives. They don't have any way out. There is nothing left for them. And, of course as the years march by, people begin to feel this more and more when it should be just the other way around. They should begin to feel freer to explore a little further, to take a few things whether they are job-related or not.

I like the ads I'm seeing now that are aimed at older women that say, "You can do this, you can do that. You raised a family all of those years. Now go out and put your talents to work." So many times people feel that life is over when it should be just beginning. They have so much more that they can do. But they are not, on a broad basis, told this. What a waste.

Betty realized early on that her age and family obligations limited the time she could devote to college. She also knew that the conventional college experience geared for eighteen-year-olds was unsuitable for her in many respects. She discovered that adult learners want to question as they learn. This means that they require a more personal forum for their classroom. She also realized, as Peg pointed out earlier, that adult learners have a great deal to offer in class from their own experience and learning. Her advisor opened an avenue for Betty to receive due credit for her experience by enrolling her in the Educational Assessment Course.

Betty also pinpointed another extremely important characteristic of adult-friendly colleges™: flexibility. By flexibility, she meant not only offering logistical leeway for busy adults who must take care of the mundane details of buying books, attending classes, and other such matters around the demands of busy work and home schedules. She also meant that colleges must be open to the varying philosophies that adults have developed outside of college, including different ways of effective learning.

Betty appreciated and enjoyed her experience at the Community College of Vermont because of the openness of the school, its recognition that adults are resources for learning as well as learners, and its willingness to give credit for learning achieved outside of the college. Without actually characterizing it as such, Betty had identified the key element of an adult-friendly college™: a cooperative learning environment.

Raye Kraft

Raye Kraft went through a similar experience at Metro State in St. Paul, Minnesota. Her observations and comments add more vital elements to the formula that makes up a college responsive to the needs of adults. An energetic woman who radiates commitment and ideas, it's hard to imagine Raye as the passive and unhappy person she describes herself as being in the past. As she rode through her early years, however, she was aware of events, of stress. But, she was not aware of the changes she was experiencing.

"When I went to Metro, they took my credits, gave me credit for learning I had done outside of school, and let me do my project. They gave me a lot of latitude and

Your Hidden Credentials

flexibility to do it. . . . If I'd gone back to the U and restarted, I wouldn't have been able to stand it. It would have been just studying and taking courses for three and a half years. It would have turned me off. I wanted to do something."

I think there were two major forces that shaped my life. One was the kind of training I had at Beth Israel Hospital in Boston, which was an excellent school of nursing. It wasn't an academic program only, it was competence-based. You were on the wards, you had to make your decisions, you had to organize right from the beginning, you had to be creative and imaginative. I have to believe that this kind of training, which was not sitting in a university for four years and learning problem-solving in the abstract way, prepared me to deal with the world.

The other factor was marrying my husband. It was tremendously significant because I came from a provincial, first-generation family. They came from the old country, settled in Boston, and had no vision of getting out of Boston or of doing anything else. Not imaginative. I married someone who told me he was going to be in Europe. He was finishing his Ph.D. at MIT at the time. Although his mother had been raised in the same town as my father, that's where the similarity ended.

He had been raised in California. He moved there after he was seven and had been exposed to a lot of different things. He was Europe, Danish furniture, and stainless flatware, and I was French provincial, knick-knacks, and doodads. He introduced me to Eames chairs, Picasso, and T.S. Eliot. Part of me was excited, but part of me was afraid and fought it, wanted to get back to those roots. The big shock came when he meant it and we moved out of Boston to Pennsylvania, where I went to school at Penn State for a couple of years; then, to Utah, where I worked at Salt Lake County General Hospital; to Oxford, England; and then to Yugoslavia. All these things had very definite bearings on what happened to me.

Then, two things happened. When I was in Yugoslavia in 1970, I saw working women. I saw that it wasn't so important to cook your own food, or bake your own cakes every time you had company; that no one really cared if your laundry looked great all the time or if you had a garden. I met women who wouldn't have called themselves liberated, but who were doing first things first, working for economic necessity. They worked and they didn't apologize. They didn't say, "Gee I didn't have time to bake." I said, "My God, you know, I can go to work. I don't have to be frustrated because I'm a lousy housekeeper and I'm never getting things clean enough or getting things right. I don't have to have the best lunches in town."

At that time I was thinking of putting an end to the marriage because I didn't know who I was, and I was putting it all on him. I came back here and got a job. When people asked me if I wasn't going to have coffee parties and be around I said, "No, I'm going to go to work." The Mrs. Doctor Professor thing was over.

I went to work at the Jewish Home for the Aged and I got very excited about the whole aging field. At first I was content to be working again, with the old people. And, as I got more and more into the field, I kept getting more turned on. But, I never had any vision of being anything beyond a nursing-home nurse.

After a while, I got into difficulties with the administrator. I tried to tell him how to run things. I thought he was too interested in the building and not enough in people. I thought we should involve the children of our patients and do seminars on "You and Your Aging Parents" and that kind of thing. He didn't want to hear it. He told me that I was administratively undisciplined and to get my creativity in bed and be a good nurse and pass my pills. We came to blows over it and I lost my job. It was a terrible thing to happen to me.

I was devastated. Here I had finally gotten a job, started an in-service program, set up a central supply unit, set up a disaster plan, and then he said that to me. It was just terrible. I thought, "Oh my God, I'm useless." So, there I was. I was no longer happy with the marriage, I was no longer happy with Mrs. Doctor Professor, I didn't want to be a housewife, and I was already sensing that I was no longer Boston.

So, I went back to the University of Minnesota and I said, "Look, I have credits from Penn State, two year's worth of liberal arts credits. I almost got a degree in public health nursing at the University, which I started but gave up to adopt a child when we first got here." They told me about Metro State. They said, "There is this new university that opened up, it's just for people like you."

"Some people need that traditional education, but it just isn't as good for this type of thing. We hire a lot of people from nontraditional education, by the way. They are more flexible, they are more imaginative, and they are more fun."

Now, I was sufficiently into this University of Minnesota professor/MIT Ph.D. kind of thing to doubt the whole idea of a college without a campus for adults. About that time, the boss I have here called me up and asked me to come for a session on what to do for the aged. I outlined what we have today, which we called then the Embryonic Aging Department. I said, "Okay, I'll develop a program. I'll develop health assessment programs." My boss said, "You know we can't fund you here because you don't have a degree; you are not an M.S.W." I said, "Okay, I'll do it as a student. If you will let me, I'll see if I can get credit at Metro State, and I'll set up an aging program for you."

So, I went to Metro State. It was this funny little place over a pharmacy, or over St. Paul's Book and Stationery, or Walgreen's or something, and I felt terrible. I thought, "What am I doing going to this crapo little place? It's not accredited or anything." But then I said, "Oh well, what have I got to lose? I want to set the program up anyway." When I got there I said, "I'm supposed to set up my own program. I'm going to develop an aging program. I'll do this, this, this, and this. I'm going to implement it and I'm going to evaluate it. How many credits will you give me for that?"

They took my Penn State credits, my nursing credits, and my Minnesota credits. They said, "Okay." I said, "Furthermore, I'm going to be 40, so I want to be done by June." This was like the July before. And I did it. I sat down in July and outlined the program. I applied for Title

III money, got it, started the program, and won an award for being the most innovative program in the country coming out of a Jewish agency. It's still the model program for Minnesota. It has grown, it is solid, it has a great reputation, and here I am. I am no longer Mrs. Doctor Professor. I'm Raye Kraft, everybody knows me, and they know the program.

When I went to Metro, they took my credits, gave me credit for learning I had done outside of school, and let me do my project. They gave me a lot of latitude and flexibility to do it, and that was very important. I wasn't getting paid to do this, and I had to have some way of saying, "I'm doing this." If I had gone back to the U and restarted, I wouldn't have been able to stand it. It would have been just studying and taking courses and exams for three and a half years. It would have turned me off. I wanted to do something. Along the way, I learned to reflect, to integrate, to analyze what was happening in my life.

Some people need that traditional education, but it just isn't as good for this type of thing. We hire a lot of people from nontraditional education, by the way. They are more flexible, they are more imaginative, and they are more fun.

It took Raye years and enormous effort to capitalize on the insight that her experiences in Yugoslavia and later losing her job gave her. Even though ". . . the Mrs. Doctor Professor thing was over . . . ," Raye encountered all of the obstacles posed by a society that is not adult-friendly.

- She doubted the value of her knowledge, skills, and abilities.
- She was frustrated by her own status concerns; her sense of what was an education and what was not.
- She lacked the proper degree or credential to do what she wanted to do.
- Her employer failed to understand and use her personal learning: the skills and abilities that she brought with her.
- She couldn't find a college that responded to her learning needs.

Raye was almost closed out. Finally, she found a working-learning situation that met her needs and built on her existing experience and knowledge. A college that respected the learning she had accomplished previously broke the logjam. This college was willing to build a current college program around her field activity.

Raye's involvement at Metro State illustrates even more clearly the importance of a college's willingness to both qualify and quantify an individual's personal learning according to her needs outside of college. Metro State was adult-friendly not only in terms of logistics and philosophy but also in its willingness to create a curriculum that centered on Raye's needs and was under her direction. The critical tools they used to do this were a documented assessment of what Raye had learned outside of the formal educational arena and a "project-based" approach to curriculum that allowed her to build her own program within college guidelines. Using these two devices, Raye was able to present her work successfully to the organizations that would implement it. Everybody was a winner in the process.

Both Betty and Raye knew from the outset that they possessed valuable personal learning and that one of the things they really needed from an adult-friendly college℠ was documentation they could use in the work place to establish their credits in advance. Phil Barrett's interest in Marylhurst ran in a similar vein. But, in his discovery of a deep well of forgotten personal learning, he brought to light yet another characteristic of adult-friendly colleges℠.

Phil Barrett

Phil Barrett was juggling several responsibilities with work and his role as husband and parent. But, he knew that if he was going to be promoted to his boss's job when the time came, a bachelor's degree would be a distinct advantage even though he had years of experience and thousands of hours invested in management training. Phil needed to find a college that would enable him to learn during the evenings and on weekends, and would give him credit for learning that he had done on the job over the years. Marylhurst, an institution near Portland, Oregon, designed to meet the needs of adults, offered these characteristics to Phil.

"I found out that the college . . . would consider giving you credit for learning you had done outside of school. . . . The program has made a big difference

for me beyond getting the credit. As I got more into the program, I found out that I knew more than I thought I did. . . . I think we lose knowledge because we aren't challenged to use it."

I think I'm a very structured person, although my psychological and sociological background indicates that a structured person is usually an engineer or a mathematician. I'm not that type of a person, but I tend to be very structured. Currently, I am the Deputy Commissioner of Public Safety, which is a real classy term that means I'm director of Police Services here. The population is about 35,000 and I have about 50 people under me.

I graduated from high school, went to work at a couple of jobs, and got into police work in 1964 as a reserve officer. I came to Beaverton as a patrolman in 1967. After a year and a half, I was promoted to sergeant and assigned to the division of detectives. As a supervisor, I was involved in setting up the crime prevention program. I also served as a training officer and was promoted up through the ranks of lieutenant and captain to deputy administrator beginning in 1975. I have an associate degree from Portland Community College and I'm a graduate of the FBI Academy in Quantico, Virginia. I've had about 1400 hours of noncredited police and technical training.

In August of 1980, I was promoted to my present position. This was a big change for me, and the thing that had the biggest impact on me there was that I was in charge of people I used to work with.

My boss is talking about retiring in a few years and I want to sit in his chair. His job description requires a bachelor's degree. From an educational standpoint, I received my associate degree in 1973. Since then, I've spent all my time taking the classes that were convenient, one-week seminars and one-day work shops. I've accumulated all this time and experience and never got any credit for it.

With the age of my children and their activities, going back to school was too much of a problem. I knew how hard I worked just for the associate degree. Now I'm at the point of saying "Hey, I need the paper, but more important, I want the satisfaction of being able to say I've got a college degree."

"Experience alone isn't enough. It's seeing what you learned and knowing how you've changed that matters."

I found out that the college here at Marylhurst would consider giving you credit for learning you had done outside of school, so I looked into it and saw what was available. The program has made a big difference for me beyond getting the credit. As I got more into the program, I found out that I knew more than I thought I did. I started listing all the things I had done in my life, and I suddenly realized that I had done a lot more than I thought that I had. The writing helped me remember, and it brought my learning all back up to the surface. I think we lose knowledge because we aren't challenged to use it. Much of what I really know is in the back of my mind. It just needed to surface.

I just did a paper on planning that had to do with PERT and MBO. I got looking at it, reading up a little on PERT and critical-path management. These are things we've been doing for years, but we never put a label on it. I didn't know it had a name.

I'm not the type of person who believes that anybody owes me anything. If I get something, I want to feel that I've earned it. The assessment program is all about what you have learned away from school. It's not just what you have done, or experienced, but what you learned as a result. That's the important part. They won't give me credit for just being there. So, I have to be able to show that I learned it. I believe that if they don't give me credit in an area where I'm asking for it, that I'm getting a fair shake.

They've got high standards. If the faculty accepts all my papers I will only have to take a few courses along with the credit that I transferred, and the credit they are awarding me for my learning outside school.

I've been learning how to see the difference between something I did and how I changed as a result. I thought about how I used to react to a certain type of supervision from my boss when I started work. Then, when I got up to be a supervisor, I observed others' management styles and how I reacted to the task. I was able to see a change in myself. I could say, "Hey, I can see where I went from being this type of person to being this type." Or, I could see when I started thinking about my responsibilities as a manager rather than just being a manager. I realized what it really meant and what I had to do.

My philosophy about adult education is, Why should I sit in class for something that I've already had more education in than what I'll get in that class? My employer is paying for this and I was able to show him that I could save up to 50 percent by going through this program. This is a college that is geared to adult education. You can see that by their scheduling. There are no day classes and that means that they are building their classes for the convenience of the learner.

When you talk about regular-age college students, we are talking about another group of people. I could not go and sit in one of those classes because the instruction wouldn't be geared to my level. If they did that, the kids wouldn't understand it. It would be a waste of my time to sit in the class. Number one, I would lose interest and number two, I couldn't contribute because, if I did, they wouldn't understand. The instructor might, but the kids wouldn't.

Often, I think we lose our learning because we aren't challenged to use it. People may be very skilled in one area and if, for some reason—well, they get promoted because they do well and pretty soon they're promoted beyond their ability. Then, the boss, if he isn't careful, can end up with an organization that is run by a bunch of idiots in the jobs they are doing who were very skilled at the jobs that they left.

Then, again, we get placed in situations where there is no incentive, no reason to use what you've learned before because nobody cares and nobody is going to let you use it in the first place. Those are two big reasons why people bury their learning and lose it at work.

Phil Barrett supplemented his existing credentials at Marylhurst, but perhaps more importantly, he discovered another great store of personal learning that he had forgotten about until that

Your Hidden Credentials

time. Phil identified an extremely important function of adult-friendly colleges℠, that of soliciting adult learners to explore their own body of personal knowledge.

In spite of Phil's successes in the past, he suffered from the disadvantage of adults that we learned about at the beginning of the book. Phil was unaware of a large body of skills he'd learned informally, and therefore he was not challenged to use them. This omission on his part proved to be detrimental to his growth as well as to society through the loss of his unused skills.

The ideal adult-friendly college℠ reaches out to adults who are unaware of their learning and invites them to discover it. This makes both pragmatic and educational sense for the college. They will fill their classrooms, but they will also help learners increase their opportunities while offering society a more dynamic population as a resource. In this way, colleges would return to the usefulness for which they were originally intended as institutions.

Betty, Raye, and Phil, as well as the other adult learners profiled here have told us what is required of an adult-friendly college℠:

- Classes and support services that are suited to the needs of adults and available at times and in places that are compatible with adult learning patterns and student needs.
- Structured assessment programs to document and credit personal learning done outside of college.
- Flexibility and openness about learning philosophies that are student-centered, including new concentrations, fields of study, and program combinations.
- Willingness to co-design curricula and programs to meet the needs of the adult learner in the work place.
- Cooperative teaching/learning environments in which students' knowledge is respected and affords them opportunities to teach what they know and what they learn, acknowledging that teachers can learn from adult students' experience.
- Flexible residency and participation patterns.
- Active promotion and pursuit of personal learning throughout the community of adults—a challenge to adults to discover their personal learning, their hidden credentials.

These characteristics exemplify how an adult-friendly college™ should operate, and you can consider them part of the criteria for seeking your own local adult-friendly learning situation.

But, even as colleges become more adult-friendly in terms of their policies and programs, there is another pressing problem: the relevance of their courses and curriculum to the work places where their adult students live. Traditional college courses generally fail to teach the essence of what it takes to be a successful worker with appropriate skills. The final step, then, must be to forge a link between the worlds of work and education to focus on the person rather than the needs of the respective systems.

One of the prime reasons that colleges have adopted their gate-keeper role is the incentive they have to offer students: a good job through the degree. Business, with its emphasis on credentialism, contributes to the frustration of adult learners' use of personal learning in the marketplace. Connecting adult-friendly colleges™ to the world of work is the beginning of creating an adult-friendly society, an ideal we'll explore in the next chapter. In such a society, the personal learning as well as the hidden credentials of adults would be valued as dynamic personal and social assets.

6.

Bringing Two Worlds Together

The worlds of work and education need each other greatly, but they have never had very much to do with each other. For years these two worlds have lived an uneasy co-existence, eyeing each other suspiciously across a chasm of differing world views, separate agendas, and distinctive cultures. As a consequence, there have been strongly mixed signals about the relationship between education and work in our society. On the one hand, we have claimed a close connection between education and work and success in the marketplace. But, the reality experienced by many people has been very different.

For example, we developed vocational education in the early 1900's to prepare students for the work place. But vocational education also revealed a dark, double standard. All too often school advisors diverted poorer and less academically productive students away from regular classrooms into vocational programs. Through this practice, vocational programs became educational dumping grounds to the lower levels of the work place instead of a launching pad to better jobs. By exception, more successful students were expected to take part in other programs, both educationally and economically.

To a certain extent, the same double standard exists on the college level today. Two new policies have radically changed post-secondary education since World War II: financial aid and the creation of community colleges. The creation of need-based financial aid scholarships through the Veterans Administration and the Office of Education have given millions of veterans and working people the means to go to college for the first time. These new financial aid programs were paralleled by a meteoric increase in the number of community colleges throughout our country, giving the new students a welcome place to go. Despite these extraordinary policy developments, however, many of the new students could not translate their new educational opportunity into significant career or educational advancement.

Community colleges have had a difficult time gaining the respect of the rest of higher education. While they are hailed as a breakthrough in making college available to working people and part-time students, there has also been a second message. If community colleges were going to be work-oriented and job-related, then like vocational schools in the past, they weren't going to have much to do with the academic mainstream. Those doors were closed and, instead of being gateways to further education and better work, community colleges became corrals. Their students were segregated from the academic world by institutional policies and standards that devalued the programs and the work accomplished at community colleges.

The more community colleges have done things differently in terms of their students, their policies, and their programs, the more segregated they have become from the academic hierarchy of post-secondary education. The community college dilemma characterizes an education world that is at odds with itself and uncomfortable with direct connections to the world of work. Everyone loses in the process. Students lose the access to college and employment that they want so much. Colleges lose students before they have achieved their potential or their educational goals. And, the marketplace loses the benefit of their unrealized capacity in the work force.

Historically, the economic consequences of this segregation have been minimal because of the seemingly limitless supply of young people entering the work force. For the last forty years employers have been able to choose from an expanding pool of people entering the work force. They were able to skim the best-educated prospects according to their needs, leaving the others to fend for themselves. Job turnover was no problem because managers could always find replacements. Employers did not need to pay close attention to, or invest heavily in school programs that were work-oriented or in job-based training for their workers.

Separated from the traditional academic institutions and undervalued outside of them, work/education programs and the people they serve traditionally have been held in low esteem by educators and employers alike. To date, both worlds have been able to afford co-existence thus avoiding the hard work of understanding each other and joining forces in cooperation.

But this is no longer the case. Continuation of this segregation by choice will have catastrophic consequences for our society and for our economy. As I said in Chapter Five, our population is "squaring" as the birthplace declines and the median age increases. One significant side effect of this extraordinary demographic shift is that those people who are working in 1987 will comprise over 80 percent of our work force in the year 2000. With fewer young people entering the work force, it is projected that the number of skilled jobs will expand faster than the number of workers trained from traditional sources to fill them.

Even as traditional manufacturing jobs give way to human and financial service industry jobs in our economy, the economy will become more knowledge-based, not less so. In addition to those shifts and the demands of the emerging service industry, our manufacturing sector will continue to do what it has always done best: invent new processes and create new techniques. Robotics will replace some workers permanently. There will be some dislocation. As painful as it is, jobs will change and people will move. But there will be good jobs for people who can learn throughout life, modifying their skills and abilities. The need for these people will increase. And the need for colleges and employers who can both foster and recognize that learning will increase also.

Job expansion is going to occur. If we don't learn how to retool our work force for changing jobs and how to capitalize on the hidden credentials of our adult population, we will be forced either to import workers or export jobs in the future. This loss of economic opportunity would have a severe impact on our society, creating an institutionalized economic underclass out of reach from education, employment, or civic productivity.

Simply put, the walls between education and work must come down. We can no longer under-utilize a work force with hidden credentials and untapped capacity while relying for their replacements on college and high school graduates with formal credentials and no experience on the job. For both social and economic reasons, we must make maximum use of our continuing adult work force, understanding that their skills and behaviors are a precious natural resource that should be developed and used.

Education and Work: Making the Connection

As Peg Moore noted so bluntly, many people with college degrees are not really prepared to work at jobs at all, in spite of having the right credentials. College courses often do not relate to real on-the-job needs, and many students enter the work force bringing student attitudes and college skills to the world of service and production. They are expected somehow to transform themselves overnight from being effective participants in the collegiate world to being constructive contributors in the world of work. This is a difficult transition to make under the best of circumstances. It is even more difficult when the courses taken at college do not relate to the real demands of the work place in general or the actual job in particular.

Given the demographic changes in our population, our historic failure to effectively connect work and education, and the rapidly changing nature of work, we must avoid the natural impulse to make college courses somehow "more relevant" to the work place. That is simply not the challenge we face. If we continue to search for relevance in courses that have been designed by colleges to prepare students for the world of work, we will fail. We should remember instead the stories of Ray Miller and Betty Hill, focus our attention on the needs and abilities of the experienced individuals in our midst, and help them establish a human connection between the cultures of education and work. This kind of relevance, more personal and dynamic, will serve everyone better, institution and individual alike.

Adults come with specific needs that have been determined by the work place, personal learning, and experience. The challenge,

Your Hidden Credentials

then, is to create relevance through education programs that are more responsive to you and your needs, more respectful of your capabilities, and more integrated with the work place. Raye Kraft developed this kind of a program at Metropolitan State when she designed and implemented an innovative aging program as the core of her college "studies." Raye's project was useful to her employer, it satisfied college requirements, and it was in tune with her interests. She was making a contribution to society while she was learning.

By its very nature, relevance must be mutual—and the mutuality must include the person, not simply the systems of work and education. With this new attitude, educators and employers will be more diagnostic and responsive, less prescriptive and predictive. They will focus less on new courses and more on new attitudes, professional relationships, and services needed by adults with significant experience and personal learning. As Martha Lucenti remarked, respect for personal learning and hidden credentials will be a key component in establishing desirable connections between people and their systems of work and education.

If we achieve these things, the needs of our colleges for students and mission, and the demands of the work place for an intellectually flexible work force will all join commonly together. Since adults currently make up almost 50 percent of the student population as well as nearly 100 percent of the worker population, the practical wisdom of drawing on their hidden credentials both as a base for continuing learning and for promotion at work is clear. Business already spends more money and has more programs for adults than our colleges and universities. The challenge now is to spend that money on programs that are appropriate to emerging needs.

New Approaches to Education for Work
We have an adult population and work force that is high in experience and low in credentials, well-qualified yet under-utilized in an economic and social environment that desperately needs their contributions. The question is, what can our colleges and employers do to help adults update, recognize, and benefit from their hidden credentials throughout life? In order to do this we and

they must change some of our most basic notions about teaching and learning. If the skill requirements for jobs continue to change as rapidly as they have been, skill attainment and upgrading will be done more productively and effectively on the job, not in a school. Correspondingly, as the cost of technology and machinery increases, it will become less attractive to duplicate the capital investments of the private sector for education programs. The public purse is going to be less and less able to support the cutting edge of technology in our schools.

The worlds of education and work must form a partnership for the ongoing education and training of adults. At the base of this partnership must be a keen realization that neither world can do the job better alone. Such cooperative agreements should include the following common elements:

- A system for combining working and learning at the job site.
- A system for assessing and updating a person's hidden credentials on a regular basis.

Perhaps the most ambitious and comprehensive example of this kind of cooperation is the College and University Options Program (CUOP) of the United Auto Workers-Ford National Development and Training Center. This consortium consists of a series of regional networks of "partner colleges" that work cooperatively through regional councils to broker services to workers. There are pilot programs in eight urban centers in the Midwest offering improved access and full service delivery to participating learners. UAW-Ford prepays the tuition to participating colleges and their workers are enrolled with full rights and privileges.

The consortium is offering full college services for UAW members on the job and in the union hall. Services include an assessment of their hidden credentials for academic credit, financial aid, other counseling and support services, and curriculum designed to be responsive both to their work and personal needs as well as the standards of the participating colleges.

Everybody wins. Management gets an upgraded and retrained work force. The UAW provides an important benefit and increased job security to its members. By tapping into a new market, the colleges enroll students they wouldn't otherwise see, and the workers get recognition of their personal learning and hidden credentials as well as the power of a pertinent education.

Other new approaches should include:

- An Experience Factor℠ rating system, to quantify personal learning and hidden credentials for the record.
- The creation of an Educational Passport that will allow people to take a record of their hidden credentials wherever they go.

At the heart of these proposals is the idea that every person has an Experience Factor℠, a value that can be attached to their personal learning and hidden credentials. Your Experience Factor℠ then could be assessed and described in terms that would be widely accepted, much the way in which grades and credits are represented on transcripts in the educational world. The key to developing a widely trusted and reliable Experience-Factor℠ rating system depends upon using a standard approach and widely accepted principals of good practice for the assessment and recognition of your hidden credentials. The Council for Adult and Experiential Learning (CAEL) has developed both the protocol, the institutional network, and the business and industry linkages to coordinate such a national service.

The next question is how to communicate your Experience Factor℠. The answer is with an Educational Passport, an up-to-date record of your hidden credentials and your formal educational experiences, that will be timely, secure, reliable, and flexible. An Educational Passport would list all of your personal learning and hidden credentials that had been assessed as well as any formal education you have done and other significant events in your life. The Passport would be organized using generally accepted categories, terms, and descriptions of learning drawn from current college transcripting practices and CAEL's protocol for good practice in assessing personal learning. An organization such as the College Board or the American College Testing Service could manage this task.

Using common language and storing the information, either in computer-accessible form or on microfiche, you could carry a validated up-to-date copy of your Passport or have immediate access to it either through a computer phone hook-up or by requesting a validated copy through the mail.

With an Educational Passport, adults like Connie Yu Naylor and Bob DePrato would possess the power to connect the worlds of education and work on their own terms. They could use Passport information with employers for the purposes of job placement and upgrading. They could also use it as a base for further personal learning or formal educational planning with colleges. Finally, the Educational Passport would be a common and reliable record of your learning history.

Recognition of your hidden credentials by both the business and college sectors is a key ingredient in this whole process. As employers learn that respect for personal learning and hidden credentials improves worker morale, which helps to keep workers at their company and also increases productivity and flexibility, they will insist on recognizing and using your hidden credentials as an asset and a resource. The economic costs associated with ignoring your hidden credentials, when combined with the human costs that occur when hidden credentials are ignored and the enormous social costs of a permanently institutionalized economic underclass are simply too great for our society or our economy to continually absorb.

In a world that values your hidden credentials, the Experience Factor and the Educational Passport will reflect capacity and knowledge, not simply your formal educational attainment. Colleges will be places where learning is recognized and valued, not just where instruction is provided. The work place will become a place where skill growth and behavior change are recognized and rewarded and where the intelligent participation of workers is respected and encouraged.

Colleges and employers must become more sensitive with their structures and their policies to your hidden credentials. But the linchpin to this future lies with your own recognition of personal learning and hidden credentials. Only you can decide to begin the process of recognizing your personal learning and hidden credentials. As individuals, we can begin to create the future we need by becoming aware of our personal learning and by respecting the skills, values, and attitudes that we have learned. Then we can respect the people we have become.

Your Hidden Credentials

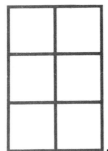

Appendix 1

Exercises to Identify Personal Learning

- Life Experiences List
- Autobiographical Exercise
- Significant Experience List

These exercises are user-oriented. They will help you get started thinking about your personal learning and hidden credentials. It is important to remember that they neither match nor duplicate either the rigor or the extensiveness of formal assessment exercises offered by adult-friendly colleges™ for credit.

You will need paper and a pencil to do the exercises.

Life Experiences List

The purpose of this exercise is to give you the chance to list all the major experiences of your life so far. This will provide you with "raw data" for later analysis.

Begin by listing when you first left school or went out on your own, and continue right down to the present. Any experience may qualify, but if you have difficulty deciding, think of those lasting for at least a month. Some areas to consider are: jobs, volunteer work, special training, raising children, travel, good books read, hobbies, personal encounters, or formal education.

On the next page is an example of what a list might look like, in part:

Dates	Experience	Comments
Summers and 7/58–9/60	Worked with pulp cutting crew and lumber company	Learned how hard it is to make a living!
10/60–12/63	U.S. Army Signal Corps	Went in as a private and came out a sergeant.
1/64 —	Fell in love and married my wife	No comment.
1/64–3/64	Took course in real estate brokering and law; passed examination	I think this was when I began to want to get someplace in life.
1/64–present	Joined Westmont Methodist Church and began working with the aged.	This started me thinking about other people for once!
4/64–present	Began selling real estate on the side; this eventually developed into my own small firm.	
8/68	Had our first child; read Dr. Spock cover to cover and everything else we could find.	This had to be one of the most important things that ever happened to me.
10/70	Had our second child; didn't read a thing.	Ditto
5/67–7/67	Went to Mexico with church group and stayed on later.	Gave us a chance to practice our Spanish; awful food, but nice people

Here is a blank chart for you to use.

Dates	Experiences	Comments

Autobiographical Approach

Some people don't know what they know until they hear themselves say it. You might try this approach with the assistance of a friend or a member of your family, who could record the important points, or you might "say" it into a tape recorder. Or, you might write an account of your important experiences. As you begin, you need not be concerned about describing learning or skills. Simply describe those experiences in life that were important to you because you:

- think you learned a lot
- are pleased with what you were able to achieve
- received appreciation/recognition from other people
- expended considerable time, energy, or money
- found the experiences enjoyable
- found the experience painful
- other reasons

Significant life experiences often build on one another. Describe how one activity may have prepared for a future activity. Be sure to include experiences you believe have been significant to your personal development and learning. What you are striving for is to identify themes—short-term and long-term experiences that are significant.

Once you have completed your autobiography, in which you have described your significant life experiences, you will be ready to proceed to the next step in the process of identifying what you learned through prior experience.

In-Service Training, Company/Professional Courses, Workshops, Conferences

_____

_____

_____

_____

_____

_____

_____

_____

_____

_____

_____

_____

_____

_____

Adult Education Courses

_____

_____

_____

_____

_____

_____

_____

_____

_____

_____

_____

_____

_____

_____

Approaches to Identification of Learning

_____

_____

_____

_____

_____

_____

_____

_____

_____

_____

_____

_____

_____

_____

Skill/Competency/ Knowledge Approach

_____

_____

_____

_____

_____

_____

_____

_____

Appendix 1

_____

_____

_____

_____

_____

   If you can identify the major areas of your knowledge immediately, make a list. For example:

_____

_____

_____

_____

_____

_____

Be specific!

_____

_____

_____

_____

_____

_____

_____

_____

_____

_____

_____

Your Hidden Credentials

_____

_____

_____

Job/Experience Approach

Some people find it easier to structure things chronologically.
For example:

Job/Experience	What I Did (duties, responsibilities)	What I Learned
Hollyhock School	Taught 3rd grade Developed curriculum—"Know Your Community." Chairman, audio/visual committee	Teaching strategies Curriculum development Knowledge of use of audio-visual aids

Watch out for duplications and repetitions.

Job Experience	What I Did	What I Learned

Significant Experiences

The purpose of this exercise is to help you narrow things down to a more workable size. You are to select from among all your experiences those that seem most significant. If you have trouble deciding, consider these questions:

- Did you get a sense of accomplishment from it?
- Do you feel you learned something important?
- Did you gain new knowledge? skills? attitudes?
- Does what you learned seem to relate to your degree goals?

As you decide, jot down a short identifying phrase beside each number below. Don't worry about the order.

1. _____

2. _____

3. _____

4. _____

5. _____

6. _____

7. _____

8. _____

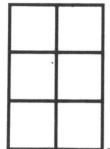

Appendix 2

Learning Resource Inventory and Planning Questions

These questions are drawn from Dr. Allen Tough's research that led to the publication of his book *The Adult's Learning Projects*. The Resource Inventory Questions are designed to help you think about the people, the things, the situations, and other devices you have used to learn over the years. You will recognize many strategies described in the book's portraits. As you recall the resources that you have used to learn, you will undoubtedly remember still more learning that you have previously overlooked.

For more information, contact:

Dr. Allen Tough
Department of Adult Education
OISE
252 Bloor St. West
Toronto, Canada MSS1V6

Resource Inventory Questions

Can you recall any other efforts to learn that were related to your home or your family? Anything related to your hobbies or recreation? Your job? Your responsibilities in various organizations, clubs, in a church or synagogue, on a committee, or some other responsibilities? Anything related to some teaching, writing, or research that you do outside of your job?

Going right back over the past 12 months, can you recall any other times that you tried to learn something by reading a book? When you read newspapers or magazines, do you read certain topics or sections because you want to *remember* the content? Have you tried to learn anything else from booklets, pamphlets, or brochures? From memos, letters, instructions, or plans? From technical or professional literature? From material from a library? From workbooks or programmed instruction? From an encyclopedia or other reference work?

Have you learned anything at all from a medical doctor? From a lawyer? From a counselor or therapist? From a financial or tax adviser? From a social worker? From a coach? From a private teacher? From a specialist or expert? From individual private lessons?

Have you learned anything from documentaries or courses on television? From TV news or some other TV programs? From radio? In a theatre?

Have you tried to learn from conversations? Or from asking questions: that is, have there been any topics or areas that you have tried to learn about from your friends or other people? Have you deliberately sought to learn by seeking out stimulating individuals? Have you tried to learn anything from your parents or your spouse? From your brother or sister? From a neighbor?

Perhaps you have learned something in some group or other? Perhaps in some meeting or discussion group? From attending a conference? From a retreat or weekend meeting? From an institute or short course or workshop? From a committee or staff meeting? From taking a course? From attending evening classes, or lectures, or a speech? From a correspondence course? From attending a club or association?

Your Hidden Credentials

Perhaps tape recordings, phonograph records, or "a language lab" helped you learn something during the past year?

Have you learned in a church or synagogue? In a college, university, or school? In some community organization? In a company or factory or office? In a government program? In an exhibition, museum, or art gallery? In some vacation program? In some extracurricular activity after school? In a club? At the "Y"? At a camp?

Can you think back to 11 months ago? Try to recall your main jobs, activities, and problems at that time. Were there any efforts to learn connected with these? How about six months ago?

Planners

There are four different sorts of learning efforts, according to who plans them. That is, a person's efforts to learn can be classified according to who was *responsible for the day-to-day planning.* We have to look at who planned or decided exactly *what and how* the person should learn at each session. For example, who decided what the person should read or hear, or what else he or she should do in order to learn?

1.   Some learners decide to attend a *group* or class or conference, and to let the group (or its leader or instructor) decide the activities and detailed subject matter from one session to the next. A group may be of any size from five persons to several hundred.

2.   In other learning efforts, the planning or deciding of the details is handled by *one person,* who helps the learner in a *one-to-one situation.* That is, there is one helper (or instructor, teacher, expert, or friend), and there is only one learner. These two persons interact face-to-face, through correspondence, or on the telephone.

Private music lessons, individual lessons from a golf pro, and being taught to drive a car by a friend are examples. Two or even three learners receiving individualized attention from one other person during the same session can be included here.

3.   In some learning projects, most of the detailed planning regarding what to learn and do at each session resides in some *object* (some nonhuman resource).

Examples of these are: a set of recordings, a series of television programs, a set of programmed instruction materials, a workbook or other printed materials, and a language lab. The learner follows the program or materials: they tell him or her what to do next.

4.   In other learning projects, *the learner* retains the major responsibility for the day-to-day planning and decision-making.

The learner may get advice from various people and use a variety of materials and resources. But he or she usually decides just what detailed subject matter to learn next, and what activities and resources to use next. Instead of turning the job of planning over to someone else, the learner makes these day-to-day decisions himself or herself.

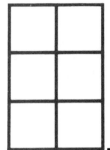

Appendix 3

Exercises to Analyze Personal Learning

- Lifeline Exercise
- Roles and Responsibilities List
- Life Experience Analysis
- Essay

While it is one thing to identify your personal learning and hidden credentials, it is quite another to begin to analyze and understand the learning that occurred as a result. These exercises will give you some simple tools with which you can begin to analyze your personal learning.

Planning for Living

Many of us act as though the future is something that happens to us rather than something we create every day. We accumulate obligations and responsibilities as we go through life, and because of this, many people tend to explain their current activities in terms of where they have been rather than where they are going. Because it is over, the past is unmanageable. Because it has not happened, the future is manageable.

The following exercises are designed to help you think about where you are, where you want to go, and the resources you have for getting there. The first is a life line where you will trace your life, taking in the past, present, and future. The second exercise asks you to focus on your present roles and responsibilities based upon that life line. In the third and fourth exercises, you will look at the future and begin to think about making plans for it.

Exercise 1: Life Line

Using the lower half of this page, draw a line to represent your life line and put a check mark on it to show where you are right now. As you proceed, try to think about the kinds of things you were doing at each stage of your life before now. (Some examples might be: schooling, jobs, travel, achievements and failures, and relationships that were/are important to you.) The line can be straight, curved, slanted, jagged, etc. Its importance is that the shape of the line represents something about how *you* think about *your* life.

Exercise 2: Roles and Responsibilities

In exploring the check mark on your life line, write *ten* different roles and/or responsibilities you now have in life. Try to list those things which are really important to your *sense of yourself,* things that, if you lost them, would make a radical difference to your identity and the meaning of *your* life.

1. _____

2. _____

3. _____

4. _____

5. _____

6. _____

7. _____

8. _____

9. _____

10. _____

Exercise 3: Life Experience Analysis

The purpose of this exercise is to help you think in detail about each of your significant experiences. This information will be important as you begin to develop your portfolio further.

1. Briefly describe the experience and what you did—or what happened to you.

   _____

   _____

2. Why was this a particularly significant experience?

   _____

   _____

3. What do you now *know* that you didn't know before?

   _____

   _____

4. What can you now *do* that you couldn't do before?

   _____

   _____

5. What do you now *believe,* or feel, that you did not before?

   _____

   _____

6. If someone else were going to do what you did in this experience, what skills, knowledge, attitudes, or qualities would they develop?

   _____

   _____

7. How do you think what you have learned applies to your future degree or personal goals.?

   _____

   _____

The General Essay Approach

The general essay describes the student's long-range plans and specific educational objectives. It describes the prior learning and shows how this learning affected your knowledge.

Questions

The following questions will help you check whether or not your description of a particular learning experience is on target.

1. Have you stated the learning activities that contributed to this knowledge?
2. When did the experience take place?
3. How long were you involved in this activity?
4. Where did this/these experiences take place?
5. How were you personally involved in these experiences?
6. What was your relationship to others and others to you? Were you responsible to others? Were others responsible to you? How many?
7. What was the name and title of the person who supervised you?
8. Have you shown that you learned and changed from the past experiences?
9. Have you described what you learned—the learning outcome of your experience?
10. Have you used specific examples or illustrations to demonstrate what you have learned?
11. Have you indicated how you can apply what you have learned?

Appendix 4

Criteria for Colleges Favorable to Adult Learners

The following categories for understanding whether a college is "adult-friendly" or not are drawn from "How to Choose: A Consumer's Guide to Understanding Colleges." The guide was written for CAEL (Council for Adult and Experiential Learning) by Sharon Hayenga, et al., in November, 1981. I selected the categories for rating and developed the simple rating scale used here to give you a tool for evaluating a specific college's readiness to serve you as an adult learner with hidden credentials. The list can serve as a check list for services as well as an evaluative tool as you try to determine if a college meets your needs.

Remember that it is an informal exercise for your own use and information, not a formal rating scale. To use the exercise, decide what criteria are important to you as an adult learner returning to college, circle the number that most accurately describes the availability of the service in question and enter that number in the blank space at the end of the line. In these rankings, a "1" means that a given service is not offered, a "3" means that the service is available but not readily accessible, and a "5" means that the service is offered and easily accessible to you as an adult learner. Then add up the right hand column and divide the total by the number of criteria you used. If the result is "3" or less, the college in question has not been organized with the adult learner in mind and is not adult-friendly. The Council for Adult and Experiential Learning also has information on adult-friendly colleges available for a small fee.

For more information contact:
CAEL
10840 Little Patunxent Pkwy.
Columbia, MD. 21044

I. Adult Learner Logistical Support Services

Parking   1 2 3 4 5 ____

Child Care   1 2 3 4 5 ____

Flexible Registration Hours and
Procedures   1 2 3 4 5 ____

Tuition Payment Options—including fee
reductions for limited services   1 2 3 4 5 ____

Class Scheduling—evenings, lunch hours,
and weekends   1 2 3 4 5 ____

Financial Aid tailored to adult needs—
part-time, day care, travel, etc.   1 2 3 4 5 ____

Public Safety during off-peak hours   1 2 3 4 5 ____

Food Services during off-peak hours   1 2 3 4 5 ____

Sub Total   _____

II. Adult Learner Academic Support Services.

Preparatory/Transitional Support Services
including skill development services   1 2 3 4 5 ____

Re-entry Student Support Group   1 2 3 4 5 ____

Orientation for Adult Learners	1 2 3 4 5	_____
Adult Services Office	1 2 3 4 5	_____
Academic Advising for Adults	1 2 3 4 5	_____
Personal Counseling for Adults	1 2 3 4 5	_____
Career Counseling for Adults	1 2 3 4 5	_____
Library Services—evenings and weekends	1 2 3 4 5	_____
Sub Total		_____

III. Academic Policies and Programs for Adult Learners

Nondegree Student Status for Returning Learners	1 2 3 4 5	_____
Flexible Residency Requirements for Adult Learners	1 2 3 4 5	_____
Assessment of Prior Experiential Learning for Academic Credit	1 2 3 4 5	_____
Specially Designed Degree Programs including independent study, community projects, and on-the-job learning	1 2 3 4 5	_____
Faculty with Experience in the Field and Trained to work with adults	1 2 3 4 5	_____
Options for Field of Concentration	1 2 3 4 5	_____
Stop In/Stop Out Policy	1 2 3 4 5	_____
Sub Total		_____
TOTAL		_____

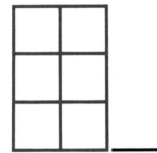

Appendix 5

A List of Colleges Favorable to Adult Learners

This is a partial and noncomprehensive list of colleges that have either begun to develop programs specifically designed for adult learners or have indicated an interest in them. It is intended as a starter list, not a compendium. Usually, the State Department of Education in your state will have more information on adult programs near you. Council for Adult and Experiential Learning also has a list.

I would like to give special notice and thanks to the colleges which allowed me to use their facilities to interview adult learners. They are: Community College of Vermont, College III at the University of Massachusetts at Boston, Bunker Hill Community College, LaGuardia Community College, St. Peter's College, Governor's State University, St. Louis University, Metropolitan State University, San Francisco State University, and Marylhurst College. Also, my thanks to the neighborhood centers and union halls in New York City and Des Moines, Iowa for their willingness to let me in.

Opportunities for Credit

ALABAMA

Alabama State University
915 South Jackson Street
Montgomery, Alabama 36195

Alexander City State Junior
College
P.O. Box 699
Alexander City, AL 35010

Athens State College
Beaty Street
Athens, AL 35611

Auburn University, Montgomery
Atlanta Highway
Montgomery, AL 36193-0401

Bessemer State Technical
College
P.O. Box 308
Bessemer, AL 35021

Gadsden State Junior College
#1 State College Boulevard
Gadsden, AL 35999-9990

George Corley Wallace State
Community College
P.O. Drawer 1049
Selma, AL 36702-1049

Hobson State Technical
College
P.O. Box 489
Thomasville, AL 36784

Huntingdon College
1500 East Fairview
Montgomery, AL 36106

Jacksonville State University
Jacksonville, AL 36265

Jefferson State Junior
College
2601 Carson Road
Birmingham, AL 35215-3098

Livingston University
Station 25, LU
Livingston, AL 35470

Miles College
P.O. Box 3800
Birmingham, AL 35208

Mobile College
P.O. Box 13220
Mobile, AL 36613

Northwest Alabama State
Junior College
Route 3 Box 77
Phil Campbell, AL 35581

Oakwood College
Huntsville, AL 35896

Opelika State Technical
College
1701 LaFayette Parkway
P.O. Box 2268
Opelika, AL 36803-2268

Patterson State Technical
College
3920 Troy Highway
Montgomery, AL 36116-2699

Shelton State Community
College
1301 15th Street East
Tuscaloosa, AL 35404

Southern Union State Junior
College
Roberts Street
Wadley, AL 36276

Troy State University
University Avenue
Troy, AL 36082

University of Alabama,
University
P.O. Box CD
University, AL 35486

University of Alabama,
Birmingham
University Station
Birmingham, AL 35294

University of North Alabama
P.O. Box 5121
Florence, AL 35632-0001

Opportunities for Credit

University of South Alabama
307 University Boulevard
Mobile, AL 36688

ALASKA
Alaska Pacific University
4101 University Drive
Anchorage, AK 99508

Anchorage Community College
2533 Providence Drive
Anchorage, AK 99508-4670

Kenai Peninsula Community
 College
Box 848
Soldotna, AK 99669

Kodiak Community College
Box 946
Kodiak, AK 99615

Sheldon Jackson College
Box 479
Sitka, AK 99835

ARIZONA
Arizona State University
Tempe, AZ 85287

Cochise College
Douglas, AZ 85607

College of Ganado
Ganado, AZ 86505

Eastern Arizona College
600 Church Street
Thatcher, AZ 85552-0769

Maricopa Technical Community
 College
108 N 40th Street
Phoenix, AZ 85034

Northern Arizona University
P.O. Box 4103
Flagstaff, AZ 86011

Northland Pioneer College
1200 Hermosa Drive
Holbrook, AZ 86025

ARKANSAS
Arkansas College
P.O. Box 2317
Batesville, AR 72503

Arkansas State University
State University, AR 72467

Harding University
Box 931, Station A
Searcy, AR 72143

Henderson State University
Arkadelphia, AR 71923

John Brown University
Siloam Springs, AR 72761

Mississippi County Community
 College
P.O. Box 1109
Blytheville, AR 72315

North Arkansas Community
 College
Pioneer Ridge
Harrison, AR 72601

Ouachita Baptist University
Box 3743
Arkadelphia, AR 71923

Southern Arkansas University
Magnolia, AR 71753

University of Arkansas,
 Fayetteville
Fayetteville, AR 72701

University of Arkansas,
 Little Rock
33rd and University Avenue
Little Rock, AR 72204

University of Arkansas for
 Medical Sciences
4301 West Markham
Little Rock, AR 72205

University of Arkansas for
 Medical Sciences, College
 of Nursing
4301 West Markham, Slot 529
Little Rock, AR 72205

Opportunities for Credit

CALIFORNIA
Allan Hancock College
800 South College Drive
Santa Maria, CA   93454

American River College
4700 College Oak Drive
Sacramento, CA   95841

Biola University
13800 Biola Avenue
La Mirada, CA   90639-0001

Cabrillo College
6500 Soquel Drive
Aptos, CA   95003

California Baptist College
8432 Magnolia Avenue
Riverside, CA   92504

California College of Arts &
  Crafts
5212 Broadway
Oakland, CA   94618

California School of
  Professional Psychology
2152 Union Street
San Francisco, CA   94123

California State College
9001 Stockdale Highway
Bakersfield, CA   93311-1099

California State Polytechnic
  University
3801 West Temple Avenue
Pomona, CA   91768

California State University,
  Chico
Chico, CA   95929

California State University,
  Northridge
18111 Nordhoff Street
Northridge, CA   91330

California State University,
  Sacramento
6000 J Street
Sacramento, CA   95819-2694

Chaffey College
5885 Haven Avenue
Alta Loma, CA   91701-3002

Chapman College
333 North Glassell Street
Orange, CA   92666

Coastline Community College
11460 Warner Avenue
Fountain Valley, CA
  92708-2597

Cogswell College
10420 Bubb Road
San Francisco, CA   95014

College of the Desert
43-500 Monterey Avenue
Palm Desert, CA   92260

Compton Community College
1111 East Artesia Boulevard
Compton, CA   90221

Consortium of the California
  State University
Academic Program Office
6300 State University Drive
Long Beach, CA   90815

Contra Costa College
2600 Mission Bell Drive
San Pablo, CA   94806

Dominican College
1520 Grand Avenue
San Rafael, CA   94901

Feather River College
Highway 70 North
Quincy, CA   95971

Fullerton College
321 East Chapman Avenue
Fullerton, CA   92634

Hartnell Community College
156 Homestead Avenue
Salinas, CA   93901

Holy Names College
3500 Mountain Boulevard
Oakland, CA   94619-9989

Opportunities for Credit

Long Beach City College
4901 East Carson Street
Long Beach, CA   90808

Los Angeles Harbor College
1111 Figueroa Place
Wilmington, CA   90744-2397

Los Angeles Pierce College
6201 Winnetka Avenue
Woodland Hills, CA   91371

Marymount Palos Verdes
 College
30800 Palos Verdes Drive East
Rancho Palos Verdes, CA 90274

Merced College
3600 M Street
Merced, CA   95340

Modesto Junior College
College Avenue
Modesto, CA   95350-9977

Moorpark College
7075 Campus Road
Moorpark, CA   93021

Mount St. Mary's College
12001 Chalon Road
Los Angeles, CA 90049

Mount San Antonio College
1100 North Grand Avenue
Walnut, CA   91789

Mount San Jacinto College
1499 North State Street
San Jacinto, CA   92383

National University
4141 Camino Del Rio South
San Diego, CA   92108

New College of California
777 Valencia Street
San Francisco, CA   94110

Orange Coast College
P.O. Box 5005
Costa Mesa, CA   92628-0120

Palomar College
1140 West Mission
San Marcos, CA   92069-1487

Sacramento City College
3835 Freeport Boulevard
Sacramento,CA 95822

Saint Mary's College
P.O. Box 785
Moraga, CA   94575

San Bernardino Valley
 College
701 South Mt. Vernon Avenue
San Bernardino, CA   92410

San Diego Mesa College
7250 Mesa College Drive
San Diego, CA   92111-4998

San Diego Miramar College
10440 Black Mountain Road
San Diego, CA   92126

San Francisco State
 University
1600 Holloway Avenue
San Francisco, CA   94132

San Joaquin Delta College
5151 Pacific Avenue
Stockton, CA   95207

Santa Ana College
17th at Bristol
Santa Ana, CA   92706

Sierra College
5000 Rocklin Road
Rocklin, CA   95677

University of California,
 Santa Barbara
Santa Barbara, CA   93106

University of LaVerne
1950 Third Street
LaVerne, CA   91750

University of the Pacific
3601 Pacific Avenue
Stockton, CA   95211

Opportunities for Credit

CALIFORNIA (cont.)
University of Redlands
1200 East Colton Avenue
Redlands, CA 92374-3755

University of San Francisco
2130 Fulton Street
San Francisco, CA 94117-1080

West Coast Christian College
6901 North Maple Avenue
Fresno, CA 93710-4599

Whittier College
13406 East Phila P.O. Box 634
Whittier, CA 90608

Woodbury University
1027 Wilshire Boulevard
Los Angeles, CA 90017

Yuba College
2088 North Beale Road
Marysville, CA 95901

COLORADO
Adams State College
Alamosa, CO 81102

Aims Community College
P.O. Box 69
Greeley, CO 80632

Arapahoe Community College
5900 South Santa Fe Drive
Littleton, CO 80120-9988

Colorado College
Colorado Springs, CO 80903

Colorado Mountain College
P.O. Box 10001
Glenwood Springs, CO 81602

Colorado School of Mines
Golden, CO 80401

Colorado State University
Fort Collins, CO 80523

Fort Lewis College
Durango, CO 81301

Front Range Community
College
3645 West 112th Avenue
Westminster, CO 80030-2199

Iliff School of Theology
2201 South University
Boulevard
Denver, CO 80210

Loretto Heights College
3001 South Federal Boulevard
Denver, CO 80236

Metropolitan State College
1006 11th Street
Denver, CO 80204

Pueblo Community College
900 West Orman Avenue
Pueblo, CO 81004

Red Rocks Community College
12600 West Sixth Avenue
Golden, CO 80401

Regis College
3539 West 50th Parkway
Denver, CO 80221

Rockmont College
180 South Garrison
Lakewood, CO 80226

University of Colorado,
School of Medicine
4200 East 9th Avenue
Denver, CO 80262

University of Northern
Colorado
Greeley, CO 80639

University of Southern
Colorado
2200 N Bonforte Boulevard
Pueblo, CO 81001

CONNECTICUT
Albertus Magnus College
700 Prospect Street
New Haven, CT 06511

Opportunities for Credit

Briarwood College
2279 Mount Vernon Road
Southington, CT   06489

Fairfield University
Fairfield, CT   06430

Greater Hartford Community
  College
61 Woodland Street
Hartford, CT   06105

Housatonic Community College
510 Barnum Avenue
Bridgeport, CT   06608

Middlesex Community College
100 Training Hill Road
Middletown, CT   06457

Mitchell College
437 Pequot Avenue
New London, CT   06320

Mohegan Community College
Mahan Drive
Norwich, CT   06360

Northwestern Connecticut
  Community College
Park Place East
Winsted, CT   06098

Norwalk Community College
333 Wilson Avenue
Norwalk, CT   06854

Quinebaug Valley Community
  College
Maple Street P.O. Box 59
Danielson, CT   06239

Quinnipiac College
Mount Carmel Avenue
Hamden, CT   06518

Sacred Heart University
P.O. Box 6460
Bridgeport, CT   06606-0460

Saint Alphonsus College
1762 Mapleton Avenue
Suffield, CT   06078

Saint Joseph College
1678 Asylum Avenue
West Hartford, CT   06117

South Central Community
  College
60 Sargent Drive
New Haven, CT   06511

Tunxis Community College
Routes 6 and 177
Farmington, CT   06032

United States Coast Guard
  Academy
New London, CT   06320

University of Hartford
200 Bloomfield Avenue
West Hartford, CT 06117-0395

Western Connecticut State
  University
181 White Street
Danbury, CT   06810

DELAWARE
Delaware State College
1200 North Dupont Highway
Dover, DE   19901

University of Delaware
Newark, DE 19716

DISTRICT OF COLUMBIA
American University
4400 Massachusetts Avenue NW
Washington, DC   20016

Gallaudet College
800 Florida Avenue NE
Washington, DC   20002

Howard University
2400 Sixth Street NW
Washintgon, DC   20059

Mount Vernon College
2100 Foxhall Road NW
Washington, DC   20007

Southeastern University
501 Eye Street SW
Washington, DC   20024

Opportunities for Credit

DISTRICT OF COLUMBIA (cont.)
Trinity College
125 Michigan Avenue NE
Washington, DC 20017

FLORIDA
Barry University
11300 NE Second Avenue
Miami, FL 33161

Bethune Cookman College
640 Second Avenue
Daytona Beach, FL 32015

Brevard Community College
Clearlake Road
Cocoa, FL 32922

Broward Community College
7200 Hollywood Boulevard
Pembroke Pines, FL 33024

Daytona Beach Community
College
P.O. Box 1111
Daytona Beach, FL 32015

Eckerd College
P.O. Box 12560
St. Petersburg, FL 33733

Edison Community College
8099 College Parkway SW
Fort Myers, FL 33907-9990

Embry-Riddle Aeronautical
University
Star Route Box 540
Bunnell, FL 32010

Flagler College
P.O. Box 1027
St. Augustine, FL 32085-1027

Florida Junior College
501 West State Street
Jacksonville, FL 32202

Florida Keys Community
College
Key West, FL 33040

Gulf Coast Community College
5230 West Highway 98
Panama City, FL 32401

Hillsborough Community
College
P.O. Box 22127
Tampa, FL 33630

Lake Sumter Community
College
5900 U S 441 South
Leesburg, FL 32748

Luther Rice Bible College &
Seminary
1050 Hendricks Avenue
Jacksonville, FL 32207

Miami-Dade Community College
11011 SW 104th Street
Miami, FL 33176

National Education Center,
Tampa Technical Institute
3920 East Hillsborough
Avenue
Tampa, FL 33610

Nova College
3301 College Avenue
Fort Lauderdale, FL 33314

Rollins College
Winter Park, FL 32789-4499

Saint Leo College
State Road 52
Saint Leo, FL 33574

Saint Petersburg Junior
College
P.O. Box 13489
St. Petersburg, FL 33733

Saint Thomas University
16400 NW 32nd Avenue
Miami, FL 33054

Southeastern College
1000 Longfellow Boulevard
Lakeland, FL 33801

Your Hidden Credentials

Opportunities for Credit

University of Central Florida
P.O. Box 25000
Orlando, FL 32816

University of Miami
University Station
Coral Gables, FL 33124

University of South Florida
4202 Fowler Avenue
Tampa, FL 33620

University of Tampa
401 West Kennedy Boulevard
Tampa, FL 33606-1490

University of West Florida
Pensacola, FL 32514-0101

Valencia Community College
P.O. Box 3028
Orlando, FL 32802

Warner Southern College
5301 U S Highway 27 South
Lake Wales, FL 33853

GEORGIA
Andrew College
College Street
Cuthbert, GA 31740

Armstrong State College
11935 Abercorn Street
Savannah, GA 31419-1997

Atlanta Christian College
2605 Ben Hill Road
East Point, GA 30344-9989

Augusta College
2500 Walton Way
Augusta, GA 30910

Berry College
Mount Berry, GA 30149

Brunswick Junior College
Altama at Fourth Street
Brunswick, GA 31523

Clayton Junior College
P.O Box 285 5900 Lee Street
Morrow, GA 30260

Dalton Junior College
Dalton, GA 30720

Dekalb Community College
955 North Indian Creek Drive
Clarkston, GA 30021

Fort Valley State College
805 State College Drive
Fort Valley, GA 31030

Georgia College
Milledgeville, GA 31061

Georgia Southern College
Statesboro, GA 30460-8033

Georgia State University
University Plaza
Atlanta, GA 30303

Macon Junior College
U S 80 and I-475
Macon, GA 31297-4899

Mercer University, Atlanta
3001 Mercer University Drive
Atlanta, GA 30341

Mercer University, Macon
1400 Coleman Avenue
Macon, GA 31207

Middle Georgia College
Cochran, GA 31014

Paine College
1235 15th Street
Augusta, GA 30910-2799

Piedmont College
Demorest, GA 30535

Reinhardt College
Waleska, GA 30183

Tift College
Tift College Drive
Forsyth, GA 31029-2318

Toccoa Falls College
Toccoa Falls, GA 30598-0068

GEORGIA (cont.)
Wesleyan College
4760 Forsyth Road
Macon, GA 31297-4299

HAWAII
Brigham Young University,
Hawaii Campus
55-220 Kulanui Street
Laie Oahu, HI 96762

Chaminade University of
Honolulu
3140 Waialae Avenue
Honolulu, HI 96816-1578

Hawaii Loa College
45-045 Kamehameha Highway
Kaneohe, HI 96744-5297

Hawaii Pacific College
1060 Bishop Street
Honolulu, HI 96813

Honolulu Community College
874 Dillingham Boulevard
Honolulu, HI 96817

Kapiolani Community College
Pensacola Street
Honolulu, HI 96814-2859

IDAHO
Boise State University
1910 University Drive
Boise, ID 83725

Idaho State University
Pocatello, ID 83209-0009

Lewis-Clark State College
8th Avenue & 6th Street
Lewiston, ID 83501

North Idaho College
1000 West Garden Avenue
Coeur D'alene, ID 83814

ILLINOIS
American Conservatory of
Music
116 South Michigan Avenue
Chicago, IL 60603

Aurora College
347 South Gladstone Avenue
Aurora, IL 60507

Belleville Area College
2500 Carlyle Road
Belleville, IL 62221

Blackburn College
Carlinville, IL 62626

Black Hawk College, East
Campus
Box 489
Kewanee, IL 61443

Chicago State University
95th Street at King Drive
Chicago, IL 60628

City Colleges of Chicago
30 East Lake Street
Chicago, IL 60601-2495

College of DuPage
22nd Street & Lambert Road
Glen Ellyn, IL 60137

College of Saint Francis
500 North Wilcox Street
Joliet, IL 60435

Columbia College
600 South Michigan Avenue
Chicago, IL 60605

De Paul University, School
for New Learning
23 East Jackson Boulevard
Chicago, IL 60604-2287

Eastern Illinois University
Charleston, IL 61920

Frontier Community College
Rural Route #1
Fairfield, IL 62837

Garrett-Evangelical
Theological Seminary
2121 Sheridan Road
Evanston, IL 60201

Opportunities for Credit

Governors State University
Park Forest South, IL 60466

Illinois Benedictine College
5700 College Road
Lisle, IL 60532-0900

Illinois Central College
East Peoria, IL 61635

Illinois College
Jacksonville, IL 62650-2299

Illinois State University
Normal, IL 61761

Illinois Valley Community
  College
2578 East 350th Road
Oglesby, IL 61348-1099

John Wood Community College
150 South 48th Street
Quincy, IL 62301-1498

Joliet Junior College
1216 Houbolt Avenue
Joliet, IL 60436-9352

Kaskaskia College
Shattuc Road
Centralia, IL 62801

Kishwaukee College
Route 38 & Malta Road
Malta, IL 60150

Lake Land College
South Route 45
Mattoon, IL 61938

Lincoln College
300 Keokuk Street
Lincoln, IL 62656

McHenry County College
Route 14 & Lucas Road
Crystal Lake, IL 60014

McKendree College
701 College Road
Lebanon, IL 62254-9990

Midstate College
244 S W Jefferson
Peoria, IL 61602

Moraine Valley Community
  College
10900 South 88th Avenue
Palos Hills, IL 60465-0937

Morton College
3801 South Central Avenue
Cicero, IL 60650

National College of
  Education, Evanston
2840 Sheridan Road
Evanston, IL 60201

National College of
  Education, Lombard
2 South 361 Glen Park Road
Lombard, IL 60148

Northeastern Illinois
  University
5500 North St. Louis Avenue
Chicago, IL 60625-4699

Northern Illinois University
De Kalb, IL 60115

Olivet Nazarene College
Kankakee, IL 60901

Prairie State College
P.O. Box 487
Chicago Heights, IL 60411

Richland Community College
2425 Federal Drive
Decatur,IL 62526

Rock Valley College
3301 North Mulford Road
Rockford, IL 61111

Roosevelt University
430 South Michigan Avenue
Chicago, IL 60605-1394

Rush University
600 South Paulina Street
Chicago, IL 60612

Opportunities for Credit

ILLINOIS (cont.)
Saint Xavier College
3700 West 103rd Street
Chicago, IL 60655

Sangamon State University
Springfield, IL 62708

Sauk Valley College
R F D No 5
Dixon, IL 61021

Sherwood Conservatory of
 Music
1014 South Michigan Avenue
Chicago, IL 60605

Southeastern Illinois College
Route 4
Harrisburg, IL 62946

Spertus College of Judaica
618 South Michigan Avenue
Chicago, IL 60605

Spoon River College
Rural Route #1
Canton, IL 61520

Thornton Community College
15800 South State Street
South Holland, IL 60473

Trinity Christian College
6601 West College Drive
Palos Heights, IL 60463

Triton College
2000 5th Avenue
River Grove, IL 60171

University of Illinois,
 Urbana-Champaign
Urbana, IL 61801

Wabash Valley College
2200 College Drive
Mount Carmel, IL 62863

Waubonsee Community College
State Route 47 at Harter Road
Sugar Grove, IL 60554

Western Illinois University
Adams Street
Macomb, IL 61455-1396

Wilbur Wright College
3400 North Austin Avenue
Chicago, IL 60634

INDIANA
Anderson College
Anderson, IN 46012-3462

Associated Mennonite
 Biblical Seminaries
3003 Benham Avenue
Elkhart, IN 46517

Ball State University
Muncie, IN 47306

Bethel College
1001 West McKinley Avenue
Mishawaka, IN 46545-5591

Butler University
4600 Sunset Avenue
Indianapolis, IN 46208

Calumet College
2400 New York Avenue
Whiting, IN 46394

DePauw University
Greencastle, IN 46135

Franklin College
501 East Monroe Street
Franklin, IN 46131-2598

Goshen College
Goshen, IN 46526-4798

Grace College
200 Seminary Drive
Winona Lake, IN 46590

Huntington College
2303 College Avenue
Huntington, IN 46750-9986

Indiana State University,
 Evansville
8600 University Boulevard
Evansville, IN 47712

Your Hidden Credentials

Opportunities for Credit

Indiana State University,
  Terre Haute
217 North 6th Street
Terre Haute, IN 47809

Indiana University,
  Bloomington
Bryan Hall
Bloomington, IN 47405

Indiana University, Southeast
4201 Grant Line Road
New Albany, IN 47150

Indiana University-Purdue
  University, Indianapolis
355 North Lansing
Indianapolis, IN 46202

Indiana Vocational Technical
  College, Central Indiana
One West 26th Street,
  P.O. Box 1763
Indianapolis, IN 46206

Indiana Vocational Technical
  College, Eastcentral
4100 Cowan Road Box 3100
Muncie, IN 47302

Indiana Vocational Technical
  College, Kokomo
1815 East Morgan Street
Kokomo, IN 46901

Indiana Vocational Technical
  College, Lafayette
3208 Ross Road Box 6299
Lafayette, IN 47903

Indiana Vocational Technical
  College, Northcentral
1534 West Sample Street
South Bend, IN 46619

Indiana Vocational Technical
  College, Northeast
3800 North Anthony Boulevard
Fort Wayne, IN 46805

Indiana Vocational Technical
  College, Southwest
3501 North 1st Avenue
Evansville, IN 47710

Indiana Vocational Technical
  College, Wabash Valley
7377 Dixie Bee Road
Terre Haute, IN 47802

Indiana Vocational Technical
  College, Whitewater
2325 Chester Boulevard
Richmond, IN 47374

Manchester College
North Manchester, IN 46962

Marian College
3200 Cold Spring Road
Indianapolis, IN 46222

Martin Center College
3553 North College Avenue
Indianapolis, IN 46205

Oakland City College
Lucretia Street
Oakland City, IN 47660-1099

Saint Mary-of-the-Woods
  College
Saint Mary-of-the-Woods, IN
  47876

Saint Meinrad College
Saint Meinrad, IN 47577

Tri-State University
South Darling Street
Angola, IN 46703-0307

Vincennes University
1002 North First Street
Vincennes, IN 47591-9986

IOWA
Briar Cliff College
3303 Rebecca Street
Sioux City, IA 51104

Clarke College
1550 Clarke Drive
Dubuque, IA 52001-9983

Eastern Iowa Community
  College
2804 Eastern Avenue
Davenport, IA 52803

Opportunities for Credit

IOWA (cont.)
Faith Baptist Bible College
1900 NW 4th Street
Ankeny, IA 50021-2198

Grand View College
1200 Grand View Avenue
Des Moines, IA 50316-1599

Indian Hills Community
 College
Grandview & North Elm
Ottumwa, IA 52501

Iowa Lakes Community College
19 South 7th Street
Estherville, IA 51334

Iowa State University
Ames, Iowa 50011

Marshalltown Community
 College
3700 South Center Street
Marshalltown, IA 50158

Marycrest College
1607 West 12th Street
Davenport, IA 52804

Mount Mercy College
1330 Elmhurst Drive N E
Cedar Rapids, IA 52402

Mount Saint Clare College
400 North Bluff Boulevard
Clinton, IA 52732

Palmer College of
 Chiropractic
1000 Brady Street
Davenport, IA 52803

Saint Ambrose College
518 West Locust Street
Davenport, IA 52803

Simpson College
Indianola, IA 50125-1299

Southeastern Community
 College
Drawer F Highway 406
West Burlington, IA 52655

University of Dubuque
2000 University Avenue
Dubuque, IA 52001

University of Northern Iowa
1222 West 27th Street
Cedar Falls, IA 50614

Waldorf College
Forest City, IA 50436

Wartburg College
222 9th Street N W
Waverly, IA 50677

Westmar College
Le Mars, IA 51031

KANSAS
Allen County Community
 College
1801 North Cottonwood
Iola, KS 66749

Cloud County Community
 College
2221 Campus Drive
Concordia, KS 66901-1002

Cowley County Community
 College
125 South Second
Arkansas City, KS 67005

Donnelly College
608 North 18th Street
Kansas City, KS 66102

Fort Hays State University
600 Park Street
Hays, KS 67601-4099

Hesston College
Hesston, KS 67062

Independence Community
 College
Brookside Drive & College
 Avenue
Independence, KS 67301-9998

Opportunities for Credit

Johnson County Community
College
12345 College at Quivira
Overland Park, KS 66210-1299

Kansas City Kansas Community
College
7250 State Avenue
Kansas City, Kansas 66112

Kansas Newman College
3100 McCormick Avenue
Wichita, KS   67213

Kansas State University
Manhattan, KS   66506

Manhattan Christian College
1407 Anderson
Manhattan, KS   66502

Mid-America Nazarene College
2030 College Way
Olathe, KS   66061-1776

Ottawa Univeristy, Kansas
City
10865 Grand View, Building 20
Overland Park, KS   66210

Ottawa University, Ottawa
10th at Cedar
Ottawa, KS   66067

Pittsburg State University
Pittsburg, KS   66762

Seward County Community
College
Box 1137
Liberal, KS   67901

Southwestern College
Winfield, KS   67156

Washburn University
17th and College
Topeka, KS   66621

Wichita State University
Wichita, KS   67208

KENTUCKY
Ashland Community College
1400 College Drive
Ashland, KY   41101

Brescia College
120 West Seventh Street
Owensboro, KY   42301

Centre College
Danville, KY   40422

Eastern Kentucky University
Richmond, KY   40475-0931

Kentucky State University
East Main Street
Frankfort, KY   40601

Kentucky Wesleyan College
Owensboro, KY   42301

Lindsey Wilson College
210 Lindsey Wilson Street
Columbia, KY   42728

Northern Kentucky University
University Drive
Highland Heights, KY   41076

Paducah Community College
P.O. Box 7380
Paducah, KY   42001-7380

Pikeville College
Sycamore Street
Pikeville, KY   41501-1194

Saint Catharine College
Saint Catharine, KY   40061

Southern Baptist Theological
Seminary
2825 Lexington Road
Louisville, KY   40280

Thomas More College
Fort Mitchell, KY   41017

Union College
College Street
Barbourville, KY   40906

Opportunities for Credit

KENTUCKY (cont.)
University of Louisville
South Third Street
Louisville, KY   40292

Western Kentucky University
Bowling Green, KY   42101

LOUISIANA
Centenary College of
  Louisiana
P.O. Box 4188
Shreveport, LA   71134-0188

Louisiana State University
  and Agricultural and
  Mechanical College
Baton Rouge, LA   70803

Louisiana State University,
  Alexandria
Alexandria, LA   71302

Louisiana State University,
  Eunice
P.O. Box 1129
Eunice, LA   70535

Louisiana State University,
  Shreveport
8515 Youree Drive
Shreveport, LA   71115

McNeese State University
4100 Ryan Street
Lake Charles, LA   70609

Nicholls State University
University Station
Thibodaux, LA   70310

Northwestern State University
Natchitoches, LA   71497

Our Lady of Holy Cross
  College
4123 Woodland Drive
New Orleans, LA   70114

Southeastern Louisiana
  University
100 West Dakota
Hammond, LA   70402

Southern University and
  Agricultural and Mechanical
  College, Baton Rouge
Baton Rouge, LA   70813

Southern University,
  New Orleans
6400 Press Drive
New Orleans, LA   70126

Xavier University of
  Louisiana
Palmetto & Pine Streets
New Orleans, LA   70125

MAINE
Andover College
901 Washington Avenue
Portland, ME   04103

Central Maine Vocational-
  Technical Institute
1250 Turner Street
Auburn, ME   04210

Eastern Maine Vocational-
  Technical Institute
354 Hogan Road
Bangor, ME   04401

Portland School of Art
97 Spring Street
Portland, ME   04101

Saint Joseph's College
North Windham, ME   04062

Southern Maine Vocational-
  Technical Institute
Fort Road
South Portland, ME   04106

University of Maine, Augusta
University Heights
Augusta, ME   04330

University of Maine, Machias
Machias, ME   04654

Univeristy of Maine, Orono
Orono, ME   04469

Opportunities for Credit

University of Southern Maine
96 Falmouth Street
Portland, ME   04103

MARYLAND
Baltimore Hebrew College
5800 Park Heights Avenue
Baltimore, MD   21215

Bowie State College
Jericho Park Road
Bowie, MD   20715

Capitol Institute of
 Technology
11301 Springfield Road
Laurel, MD   20708

Catonsville Community
 College
800 South Rolling Road
Baltimore, MD   21228

Cecil Community College
1000 North East Road
North East, MD   21901-1999

Chesapeake College
Wye Mills, MD   21679

College of Notre Dame of
 Maryland
4701 North Charles Street
Baltimore, MD   21210

Columbia Union College
7600 Flower Avenue
Takoma Park, MD   20912

Dundalk Community College
7200 Sollers Point Road
Baltimore, MD   21222-4692

Frostburg State College
Frostburg, MD   21532

Garrett Community College
Mosser Road
McHenry, MD   21541

Goucher College
Towson, MD   21204

Hagerstown Junior College
751 Robinwood Drive
Hagerstown, MD   21740-6590

Harford Community College
401 Thomas Run Road
Bel Air, MD   21014

Hood College
Rosemont Avenue
Frederick, MD   21701

Howard Community College
Little Patuxent Parkway
Columbia, MD   21044-3197

Maryland Institute College
 of Art
1300 West Mount Royal Avenue
Baltimore, MD   21217

Montgomery College
51 Mannakee Street
Rockville, MD   20850

Mount Saint Mary's College
Emmitsburg, MD   21727

Prince George's Community
 College
301 Largo Road
Largo, MD   20772

Saint Mary's College of
 Maryland
Saint Mary's City, MD   20686

Salisbury State College
Salisbury, MD   21801

United States Naval Academy
Annapolis, MD   21402

University of Baltimore
Charles at Mount Royal
Baltimore, MD   21201

University of Maryland,
 College Park
College Park, MD   20742

Opportunities for Credit

MARYLAND (cont.)
University of Maryland,
University College
University Boulevard at
Adelphi Road
College Park, MD 20742

Villa Julie College
Green Spring Valley Road
Stevenson, MD 21153

Washington College
Chestertown, MD 21620

Western Maryland College
Westminster, MD 21157-4390

MASSACHUSETTS
American International
College
1000 State Street
Springfield, MA 01109

Assumption College
500 Salisbury Street
Worcester, MA 01609-1296

Atlantic Union College
South Lancaster, MA 01561

Berkshire Christian College
Lenox, MA 01240

Berkshire Community College
West Street
Pittsfield, MA 01201

Bradford College
320 South Main Street
Bradford, MA 01830

Brandeis University
415 South Street
Waltham, MA 02154

Bridgewater State College
Bridegwater, MA 02324

Bristol Community College
777 Elsbree Street
Fall River, MA 02720-7395

Central New England College
of Technology
678 Main Street
Worcester, MA 01610

Clark University
950 Main Street
Worcester, MA 01610

College of Our Lady of the
Elms
291 Springfield Street
Chicopee, MA 01013-2839

Curry College
1071 Blue Hill Avenue
Milton, MA 02186

Fitchburg State College
160 Pearl Street
Fitchburg, MA 01420

Greenfield Community College
1 College Drive
Greenfield, MA 01301

Hampshire College
Amherst, MA 01002

Hellenic College
50 Goddard Avenue
Brookline, MA 02146

Laboure College
2120 Dorchester Avenue
Boston, MA 02124-5698

Massachusetts College of Art
621 Huntington Avenue
Boston, MA 02115

Massachusetts College of
Pharmacy and Allied Health
Services
179 Longwood Avenue
Boston, MA 02115

Middlesex Community College
Springs Road
Bedford, MA 01730

Mount Ida College
777 Dedham Street
Newton Centre, MA 02159

Your Hidden Credentials

Opportunities for Credit

Mount Wachusett Community
  College
444 Green Street
Gardner, MA   01440

New England College of
  Optometry
424 Beacon Street
Boston, MA   02115

North Adams State College
Church Street
North Adams, MA   01247

North Shore Community
  College
3 Essex Street
Beverly, MA   01915

Pine Manor College
400 Heath Street
Chestnut Hill, MA   02167

Salem State College
352 Lafayette Street
Salem, MA   01970

Simmons College
300 The Fenway
Boston, MA   02115

Simon's Rock of Bard College
Great Barrington, MA   01230

Springfield Technical
  Community College
Armory Square
Springfield, MA   01105

University of Massachusetts,
  Amherst
Amherst, MA   01003

University of Massachusetts,
  Boston
Harbor Campus
Boston, MA   02125

Wentworth Institute of
  Technology
550 Huntington Avenue
Boston, MA   02115

Western New England College
1215 Wilbraham Road
Springfield, MA   01119-2684

Wheelock College
200 The Riverway
Boston, MA   02215

Worcester State College
486 Chandler Street
Worcester, MA   01602-2597

MICHIGAN
Adrian College
Adrian, MI   49221-2575

Andrews University
Berrien Springs, MI   49104

Aquinas College
1607 Robinson Road SE
Grand Rapids, MI   49506

Baker Junior College
1110 Eldon Baker Drive
Flint, MI   48507-1986

Central Michigan University
Mount Pleasant, MI   48859

Concordia College
4090 Geddes Road
Ann Arbor, MI   48105-2797

Delta College
University Center, MI   48710

Ferris State College
Big Rapids, MI   49307

Grand Rapids Baptist College
1001 East Beltline Avenue NE
Grand Rapids, MI   49505

Grand Rapids Junior College
143 Bostwick Avenue NE
Grand Rapids, MI   49503

Jackson Community College
2111 Emmons Road
Jackson, MI   49201

MICHIGAN (cont.)
Kalamazoo College
1200 Academy Street
Kalamazoo, MI 49007

Kalamazoo Valley Community
College
6767 West O Avenue
Kalamazoo, MI 49009

Kirtland Community College
Route #4 Box 59-A
Roscommon, MI 48653-9721

Lake Superior State College
Sault Sainte Marie, MI
49783-9981

Madonna College
36600 Schoolcraft Road
Livonia, MI 48150-1173

Marygrove College
8425 West McNichols Road
Detroit, MI 48221-2599

Michigan Christian College
800 West Avon Road
Rochester, MI 48063

Michigan State University
50 Kellogg Center
East Lansing, MI 48824-1022

Monroe County Community
College
1555 S Raisinville Road
Monroe, MI 48161

Muskegon Community College
221 South Quarterline Road
Muskegon, MI 49442

Nazareth College
Nazareth, MI 49074

Northwood Institute
3225 Cook Road
Midland, MI 48640

Oakland Community College
Box 812
Bloomfield Hills, MI 48013

Oakland University
Rochester, MI 48063

Sacred Heart Seminary
College
2701 Chicago Boulevard
Detroit, MI 48206

Saint Clair County Community
College
323 Erie Street
Port Huron, MI 48060

Southwestern Michigan
College
Cherry Grove Road
Dowagiac, MI 49047

Spring Arbor College
106 Main Street
Spring Arbor, MI 49283

Suomi College
Hancock, MI 49930

University of Michigan
Ann Arbor, MI 48109

Wayne State University
Detroit, MI 48202

West Shore Community College
3000 North Stiles
Scottville, MI 49454-0277

Western Michigan University
Kalamazoo, MI 49008-3899

MINNESOTA
Anoka Ramsey Community
College
11200 Mississippi River
Boulevard NW
Coon Rapids, MN 55433

Arrowhead Community College
Region, Vermilion Community
College
1900 East Camp Street
Ely, MN 55731

Austin Community College
1600 8th Avenue NW
Austin, MN 55912

Opportunities for Credit

Bemidji State University
Bemidji, MN 56601

College of Saint Scholastica
1200 Kenwood Avenue
Duluth, MN 55811

College of Saint Teresa
Winona, MN 55987-0837

College of Saint Thomas
2115 Summit Avenue
Saint Paul, MN 55105

Concordia College,
 Saint Paul
Hamline and Marshall Avenue
Saint Paul, MN 55104

Dr. Martin Luther College
College Heights
New Ulm, MN 56073

Golden Valley Lutheran
 College
6125 Olson Highway
Minneapolis, MN 55422

Inver Hills Community
 College
8445 College Trail
Inver   Grove   Heights,   MN
55075

Lakewood Community College
3401 Century Avenue
White Bear Lake, MN 55110

Mankato State University
South Road and Ellis Avenue
Mankato, MN 56001

Metropolitan State
 University
121 Metro Square
Saint Paul, MN 55101

Minneapolis Community
 College
1501 Hennepin Avenue
Minneapolis, MN 55403

Minnesota Bible College
920 Mayowood Road SW
Rochester, MN 55902

Moorhead State University
11th Street South
Moorhead, MN 56560-9980

North Hennepin Community
 College
7411 85th Avenue North
Brooklyn Park, MN 55445

Rochester Community College
Highway 14 East
Rochester, MN 55904

Saint Cloud State University
Saint Cloud, MN 56301

Saint Mary's College
Winona, MN 55987

University of Minnesota,
 Morris
Morris, MN 56267

University of Minnesota
 Technical College, Waseca
Waseca, MN 56093

Willmar Community College
Willmar, MN 56201

Worthington Community
 College
1450 Collegeway
Worthington, MN 56187

MISSISSIPPI
Alcorn State University
Lorman, MS 39096-9998

Millsaps College
Jackson, MS 39210-0001

Mississippi Gulf Coast
 Junior College
Perkinston, MS 39573

Reformed Theological
 Seminary
5422 Clinton Boulevard
Jackson, MS 39209

Opportunities for Credit

MISSISSIPPI (cont.)
University of Mississippi
University, MS 38677

Wesley College
Florence, MS 39073-0070

William Carey College
Hattiesburg, MS 39401

MISSOURI
Avila College
11901 Wornall Road
Kansas City, MO 64145-9990

Calvary Bible College
Kansas City, MO 64147

Central Missouri State
 University
Warrensburg, MO 64093

Columbia College
10th and Rogers
Columbia, MO 65216

Cottey College
1000 West Austin
Nevada, MO 64772-1000

Crowder College
Neosho, MO 64850

Culver-Stockton College
College Hill
Canton, MO 63435-9989

East Central College
P.O. Box 529
Union, MO 63084

Evangel College
1111 North Glenstone
Springfield, MO 65802

Fontbonne College
6800 Wydown Boulevard
Saint Louis, MO 63105

Hannibal-LaGrange College
Hannibal, MO 63401

Harris-Stowe State College
3026 Laclede Avenue
Saint Louis, MO 63103

Jefferson College
Hillsboro, 63050-1000

Kemper Military School and
 College
701 Third Street
Boonville, MO 65233

Lincoln University
820 Chestnut
Jefferson City, MO 65101

Lindenwood College
First Capitol and
 Kingshighway
Saint Charles, MO 63301

Longview Community College
500 Longview Road
Lee's Summit, MO 64063

Maple Woods Community
 College
2601 NE Barry Road
Kansas City, MO 64156-1299

Maryville College
13550 Conway Road
Saint Louis, MO 63141

Missouri Western State
 College
4525 Downs Drive
Saint Joseph, MO 64507-2294

Moberly Area Junior College
College and Rollins Streets
Moberly, MO 65270

Park College
Mackay Hall
Parkville, MO 64152

Saint Louis College of
 Pharmacy
4588 Parkview Place
Saint Louis, MO 63110

Opportunities for Credit

Saint Louis Community College
5801 Wilson Avenue
Saint Louis, MO 63110

Saint Louis Community,
  Meramec
11333 Big Bend Boulevard
Kirkwood, MO 63122

Saint Louis University,
  Metropolitan College
221 North Grand Boulevard
Saint Louis, MO 63103

State Fair Community College
1900 Clarendon Road
Sedalia, MO 65301

Stephens College
Columbia, MO 65215-0001

University of Missouri,
  Columbia
Columbia, MO 65211

University of Missouri,
  Kansas City
5100 Rockhill Road
Kansas City, MO 64110

University of Missouri,
  Saint Louis
8001 Natural Bridge Road
Saint Louis, MO 63121

Webster University
470 East Lockwood
Saint Louis, MO 63119-3194

William Jewell College
Liberty, MO 64068

MONTANA
  Carroll College
  Helena, MT 59625

Dawson Community College
Box 421
Glendive, MT 59330

Eastern Montana College
1500 North 27th Street
Billings, MT 59101

Miles Community College
2715 Dickinson
Miles City, MT 59301

Rocky Mountain College
1511 Poly Drive
Billings, MT 59102-1796

NEBRASKA
  Bellevue College
  Bellevue, NE 68005-3098

Central Community College,
  Hastings
P.O. Box 1024
Hastings, NE 68901

Central Community College,
  Platte
P.O. Box 1027
Columbus, NE 68601

Central Technical Community
  College Area
P.O. Box C
Grand Island, NE 68802-0240

Chadron State College
10th and Main Streets
Chadron, NE 69337

College of Saint Mary
1901 South 72nd Street
Omaha, NE 68124

Dana College
Blair, NE 68008

Doane College
Crete, NE 68333

Hastings College
7th and Turner Avenue
Hastings, NE 68901

Mid-Plains Community
  College Area
416 North Jeffers
North Platte, NE 69101

Mid-Plains Community College
Route 4, Box 1
North Platte, NE 69101

Opportunities for Credit

NEBRASKA (cont.)
Nebraska Wesleyan University
5000 Saint Paul Street
Lincoln, NE   68504

Northeast Technical Community
  College
801 East Benjamin Avenue
P.O. Box 469
Norfolk, NE   68701

Peru State College
Peru, NE   68421

University of Nebraska,
  Lincoln
14th and R Streets
Lincoln, NE   68588

Wayne State College
Wayne, NE   68787

NEW HAMPSHIRE
Colby-Sawyer College
New London, NH   03257

Franklin Pierce College
Rindge, NH   03461

Keene State College
229 Main Street
Keene, NH   03401-4183

New Hampshire Technical
  Institute
Fan Road
Concord, NH   03301

New Hampshire Vocational-
  Technical College,
  Berlin
2020 Riverside Drive
Berlin, NH   03570-3799

New Hampshire Vocational-
  Technical College,
  Manchester
1066 Front Street
Manchester, NH   03102

New Hampshire Vocational-
  Technical College, Nashua
505 Amherst Street
Nashua, NH   03063-1092

Notre Dame College
2321 Elm Street
Manchester, NH   03104

Plymouth State College
Plymouth, NH   03264

School for Lifelong Learning
Dunlap Center
Durham, NH   03824-3545

NEW JERSEY
Brookdale Community College
Newman Springs Road
Lincroft, NJ   07738

Camden County College
P.O. Box 200
Blackwood, NJ   08012-0200

Centenary College
400 Jefferson Street
Hackettstown, NJ   07840

Cumberland County College
P.O. Box 517
Vineland NJ   08360-0517

Essex County College
303 University Avenue
Newark, NJ   07102

Georgian Court College
Lakewood, NJ   08701

Glassboro State College
Glassboro, NJ   08028

Gloucester County College
Tanyard Road-Deptford P.O.
Sewell, NJ   08080

Jersey City State College
2039 Kennedy Boulevard
Jersey City, NJ   07305-1597

Kean College of New Jersey
Morris Avenue
Union, NJ   07083

Mercer County Community
  College
1200 Old Trenton Road
Trenton, NJ   08690

Your Hidden Credentials

Opportunities for Credit

Middlesex County College
Edison, NJ   08818

Monmouth College
Cedar and Norwood Avenues
West Long Branch, NJ   07764

New Brunswick Theological
  Seminary
17 Seminary Place
New Brunswick, NJ   08901

New Jersey Institute of
  Technology
323 High Street
Newark, NJ   07102

Passaic County Community
  College
College Boulevard
Paterson, NJ   07509

Ramapo College of New Jersey
505 Ramapo Valley Road
Mahwah, NJ   07430

Saint Peter's College
2641 Kennedy Boulevard
Jersey City, NJ   07306

Salem Community College
Penns Grove, NJ   08069-2799

Somerset County College
P.O. Box 3300
Somerville, NJ   08876-1265

Thomas A. Edison State
  College
101 West State Street
Trenton, NJ   08625

Trenton State College
CN 550 Hillwood Lakes
Trenton, NJ   08625

Union County College
1033 Springfield Avenue
Cranford, NJ   07016

Upsala College
East Orange, NJ   07019

William Paterson College
300 Pompton Road
Wayne, NJ   07470

NEW MEXICO
College of Santa Fe
St. Michael's Drive
Santa Fe, NM   87501

College of the Southwest
Lovington Highway
Hobbs, NM   88240

Eastern New Mexico
  University, Portales
Portales, NM   88130

Eastern New Mexico
  University, Roswell
P.O. Box 6000
Roswell, NM   88201

New Mexico Highlands
  University
Las Vegas, NM   87701

New Mexico State
  University, Alamogordo
Box 477
Alamogordo, NM   88310

San Juan College
4601 College Boulevard
Farmington NM   87401

Western New Mexico
  University
Silver City, NM   88061

NEW YORK
Academy of Aeronautics
La Guardia Airport
Flushing, NY   11371

Adelphi University,
  University College
Garden City, NY   11530

Adirondack Community College
Glens Falls, NY   12801

Opportunities for Credit

NEW YORK (cont.)
Albany College of Pharmacy
  of Union University
106 New Scotland Avenue
Albany, NY 12208

Boricua College
3755 Broadway
New York, NY 10032

Broome Community College
P.O. Box 1017
Binghamton, NY 13902

Canisius College
2001 Main Street
Buffalo, NY 14208

Cayuga County Community
  College
Franklin Street
Auburn, NY 13021

Christ the King Seminary
711 Knox Road   Box 160
East Aurora, NY 14052

City University of New York,
  City College
Convent Avenue at 138th
  Street
New York, NY 10031

Clinton Community College
Plattsburgh, NY 12901

Colgate University
Hamilton, NY 13346

College of Insurance
101 Murray Street
New York, NY 10007

College of New Rochelle,
  School of New Resources
New Rochelle, NY 10801

Community College of the
  Finger Lakes
Lincoln Hill
Canandaigua, NY 14424

Cooper Union
41 Cooper Square
New York, NY 10003

Corning Community College
Corning, NY 14830

Daeman College
4380 Main Street
Amherst, NY 14226

Dominican College of
  Blauvelt
Western Highway
Orangeburg, NY 10962

Dowling College
Idle Hour Boulevard
Oakdale Long Island, NY
  11769

D'Youville College
320 Porter Avenue
Buffalo, NY 14201

Elizabeth Seton College
1061 North Broadway
Yonkers, NY 10701

Elmira College
Park Place
Elmira, NY 14901-2099

Erie Community College,
  City Campus
121 Ellicott Street
Buffalo, NY 14203

Erie Community College,
  North
Main Street & Youngs Road
Buffalo, NY 14221

Erie Community College,
  South
4140 Southwestern Boulevard
Orchard Park, NY 14127

Fashion Institute of
  Technology
227 West 27th Street
New York, NY 10001

Opportunities for Credit

Fulton-Montgomery Community
 College
Route 67
Johnstown, NY   12095

Hartwick College
Oneonta, NY   13820

Herkimer County Community
 College
Reservoir Road
Herkimer, NY   13350

Houghton College
Houghton, NY   14744

Iona College
New Rochelle, NY   10801-1890

Ithaca College
Ithaca, NY   14850

Jamestown Community College
525 Falconer Street
Jamestown, NY   14701

Keuka College
Keuka Park, NY   14478-0098

Kingsborough Community
 College
2001 Oriental Boulevard
Brooklyn, NY   11235

Long Island University,
 Southampton
Montauk Highway
Southampton, NY   11968

Manhattanville College
Purchase, NY   10577

Marist College
82 North Road
Poughkeepsie, NY   12601

Marymount College
Tarrytown, NY   10591

Mater Dei College
Rural 2
Ogdensburg, NY   13669

Medaille College
18 Agassiz Circle
Buffalo, NY   14214

Mohawk Valley Community
 College
1101 Sherman Drive
Utica, NY   13501-5394

Molloy College
1000 Hempstead Avenue
Rockville Centre, NY   11570

Monroe Business Institute
29 East Fordham Road
Bronx, NY   10468

Mount Saint Mary College
Newburgh, NY   12550

Nassau Community College
Stewart Avenue
Garden City, NY   11530

Nazareth College of
 Rochester
4245 East Avenue
Rochester, NY   14610

New York University
70 Washington Square South
New York, NY   10012

Onondaga Community College
Onondaga Hill Road
Syracuse, NY   13215

Orange County Community
 College
115 South Street
Middletown, NY   10940

Pace University, New York
Pace Plaza
New York, NY   10038

Paul Smith's College
Paul Smiths, NY   12970

Pratt Institute
200 Willoughby Avenue
Brooklyn, NY   11205

Opportunities for Credit

NEW YORK (cont.)
Queensborough Community
   College
Bayside
New York, NY 11364

Rensselaer Polytechnic
   Institute
Troy, NY 12181

Roberts Wesleyan College
2301 Westside Drive
Rochester, NY 14624

Rockland Community College
145 College Road
Suffern, NY 10901

Saint Francis College
180 Remsen Street
Brooklyn, NY 11201

School of Visual Arts
209 East 23rd Street
New York, NY 10010

Siena College
Loudonville, NY 12211

Skidmore College, University
   Without Walls
Saratoga Springs, NY
   12866-0851

State University of New
   York, Agricultural &
   Technical College, Alfred
Alfred, NY 14802-1196

State University of New York
   Agricultural & Technical
   College, Canton
Canton, NY 13617

State University of New York
   Agricultural & Technical
   College, Delhi
Delhi, NY 13753-1190

State University of New York
   Agricultural & Technical
   College, Farmingdale
Melville Road
Farmingdale, NY 11735

SUNY, Brockport
Brockport, NY 14420

SUNY, Buffalo
1300 Elmwood Avenue
Buffalo, NY 14222

SUNY, Empire State College
2 Union Avenue
Saratoga Springs, NY 12866

SUNY, Fredonia
Fredonia, NY 14063

SUNY, Geneseo
Geneseo, NY 14454

SUNY, New Paltz
New Paltz, NY 12561

SUNY, Old Westbury
Box 210
Old Westbury, NY 11568

SUNY, Oneonta
Oneonta, NY 13820-1361

SUNY, Oswego
Oswego, NY 13126

SUNY, Purchase
Purchase, NY 10577

SUNY, Stony Brook
Stony Brook, NY 11794

Suffolk County Community
   College, Ammerman
533 College Road
Selden, NY 11784

Trocaire College
110 Red Jacket Parkway
Buffalo, NY 14220

Ulster County Community
   College
Stone Ridge, NY 12484

University of Rochester
Rochester, NY 14627

Your Hidden Credentials

Opportunities for Credit

University of the State of
New York, Regents College
Degrees
Cultural Education Center
Albany, NY 12230

Utica College of Syracuse
University
Burrstone Road
Utica, NY 13502

NORTH CAROLINA
Belmont Abbey College
Belmont, NC 28012

Bladen Technical College
P.O. Box 1266
Dublin, NC 28332-0266

Catawba College
Salisbury, NC 28144-2488

Central Piedmont Community
College
P.O. Box 35009
Charlotte, NC 28235

Coastal Carolina Community
College
444 Western Boulevard
Jacksonville, NC 28540-6877

Craven Community College
P.O. Box 885
New Bern, NC 28560

Davidson College
Davidson, NC 28036

East Carolina University
Greenvile, NC 27834

Edgecombe Technical College
2009 West Wilson
Tarboro, NC 27886

Elizabeth City State
University
Parkview Drive
Elizabeth City, NC 27909

Fayetteville Technical
Institute
P.O. Box 35236
Fayettville, NC 28303-0236

Forsyth Technical Institute
2100 Silas Creek Parkway
Winston-Salem, NC
27103-5197

Gardner-Webb College
Boiling Springs, NC 28017

Guilford Technical Community
College
P.O. Box 309
Jamestown, NC 27282

Halifax Community College
P.O. Box 809
Weldon, NC 27890

Haywood Technical College
Freedlander Drive
Clyde, NC 28721

Isothermal Community College
P.O. Box 804
Spindale, NC 28160-0804

James Sprunt Technical
College
P.O. Box 398
Kenansville, NC 28349-0398

Lenoir-Rhyne College
Hickory, NC 28603

Mayland Technical College
P.O. Box 547
Spruce Pine, NC 28777

Meredith College
Raleigh, NC 27607-5298

Mount Olive College
Mount Olive, NC 28365

North Carolina Central
University
Durham, NC 27707

Opportunities for Credit

NORTH CAROLINA (cont.)
Pfeiffer College, Charlotte
1416 East Morehead Street
Charlotte, NC 28204

Piedmont Technical College
P.O. Box 1197
Roxboro, NC 27573

Rowan Technical College
P.O. Box 1595
Salisbury, NC 28144

Sacred Heart College
414 North Main Street
Belmont, NC 28012

Saint Andrews Presbyterian
College
Laurinburg, NC 28352

Salem College
Winston-Salem, NC 27108

Shaw University
118 East South Street
Raleigh, NC 27611

Southeastern Community
College
P.O. Box 151
Whiteville, NC 28472-0151

Southwestern Technical
College
275 Webster Road
Sylva, NC 28779

Surry Community College
South Main Street
Dobson, NC 27017-0304

University of North
Carolina, Charlotte
UNCC Station
Charlotte, NC 28223

University of North
Carolina, Greensboro
1000 Spring Garden Street
Greensboro, NC 27412-5001

University of North
Carolina, Wilmington
601 South College Road
Wilmington, NC 28403-3297

Warren Wilson College
701 Warren Wilson Road
Swannanoa, NC 28778-2099

Wayne Community College
Caller Box 8002
Goldsboro, NC 27533-8002

Western Carolina University
Cullowhee, NC 28723

Western Piedmont Community
College
1001 Burkemont Avenue
Morgantown, NC 28655-9978

Wilson County Technical
Institute
902 Herring Avenue
Wilson, NC 27893-4305

NORTH DAKOTA
Dickinson State College
Dickinson, ND 58601-4896

Lake Region Community
College
College Drive
Devils Lake, ND 58301

Mary College
Apple Creek Road
Bismarck, ND 58501

Trinity Bible College
50 6th Avenue South
Ellendale, ND 58436

Valley City State College
College Street
Valley City, ND 58072

OHIO
Antioch University
Yellow Springs, OH 45387

Ashland College
401 College Avenue
Ashland, OH 44805

Opportunities for Credit

Baldwin-Wallace College
275 Eastland Road
Berea, OH 44017

Bowling Green State
 University, Bowling Green
Bowling Green, OH 43403

Bowling Green State
 University, Firelands
 College
901 Rye Beach Road
Huron, OH 44839

Capital University
East Main Sreet
Columbus, OH 43209

Chatfield Colege
Saint Martin, OH 45118

Cincinnati Technical College
3520 Central Parkway
Cincinnati, OH 45223

Clark Technical College
570 East Leffel Lane
Springfield, OH 45505

Cleveland Institute of Music
11021 East Boulevard
Cleveland, OH 44106

College of Mount Saint
 Joseph
Mount Saint Joseph, OH
 45051

Columbus Technical Institute
P.O. Box 1609
550 East Spring Street
Columbus, OH 43216-9965

Cuyahoga Community College
700 Carnegie Avenue
Cleveland, OH 44115

Cuyahoga Community College,
 Western Campus
11000 West Pleasant Valley
 Road
Parma, OH 44130

Defiance College
Defiance, OH 43512

Denison University
Granville, OH 43023

Findlay College
1000 North Main Street
Findlay, OH 45840

Franklin University
201 South Grant Avenue
Columbus, OH 43215-5399

Heidelberg Colege
310 East Market Street
Tiffin, OH 44883

Hiram College
Hiram, OH 44234

Hocking Technical College
Route 1
Nelsonville, OH 45764-9704

Kent State University,
 Ashtabula
3325 West 13th Street
Ashtabula, OH 44004

Kent State University,
 Tuscarawas Campus
University Drive NE
New Philadelphia, OH 44663

Lake Erie College
391 West Washington Street
Painesville, OH 44077

Lakeland Community College
Mentor, OH 44060

Lima Technical College
4240 Campus Drive
Lima, OH 45804

Lorain County Community
 College
1005 North Abbe Road
Elyria, OH 44035

Lourdes College
6832 Convent Boulevard
Sylvania, OH 43560

Opportunities for Credit

OHIO (cont.)
Malone College
515 25th Street NW
Canton, OH 44709-3897

Miami University, Hamilton
1601 Peck Boulevard
Hamilton, OH 45011

Miami University, Oxford
Oxford, OH 45056

Mount Union College
1972 Clark Avenue
Alliance, OH 44601

Mount Vernon Nazarene
 College
800 Martinsburg Road
Mount Vernon, OH 43050

Muskingum Area Technical
 College
1555 Newark Road
Zanesville, OH 43701

Muskingum College
New Concord, OH 43762-1199

North Central Technical
 College
2441 Kenwood Circle
Mansfield, OH 44906

Notre Dame College
4545 College Road
Cleveland, OH 44121

Ohio Dominican College
1216 Sunbury Road
Columbus, OH 43219

Ohio State University,
 Columbus
Admissions Office, Third
 Floor, Lincon Tower
1800 Cannon Drive
Columbus, OH 43210-1358

Ohio State University, Lima
4240 Campus Drive
Lima, OH 45804

Ohio State University,
 Mansfield
1680 University Drive
Mansfield, OH 44906

Ohio State University,
 Marion
1465 Mount Vernon Avenue
Marion, OH 43302-5695

Ohio University, Athens
Athens, OH 45701

Otterbein College
Westerville, OH 43081

Owens Technical College
Oregon Road
Toledo, OH 43699

Shawnee State Community
 College
940 Second Street
Portsmouth, OH 45662

Sinclair Community College
444 West Third Street
Dayton, OH 45402-1460

Southern State Community
 College
200 Hobart Drive
Hillsboro, OH 45133

Stark Technical College
6200 Frank Avenue NW
Canton, OH 44720

Terra Technical College
1120 Cedar Street
Fremont, OH 43420

Union for Experimenting
 Colleges and Universities
632 Vine Street Suite 1010
Cincinnati, OH 45202-2407

University of Cincinnati,
 Cincinnati
Cincinnati, OH 45221

Opportunities for Credit

University of Cincinnati,
Clermont College
College Drive
Batavia, OH   45103-1785

University of Steubenville
Franciscan Way
Steubenville, OH   43952

University of Toledo
2801 West Bancroft Street
Toledo, OH   43606

University of Toledo,
Community & Technical
College
Toledo, OH   43606

Urbana University
College Way
Urbana, OH   43078-9988

Ursuline College
2550 Lander Road
Cleveland, OH   44124

Walsh College
2020 Eastern Street NW
Canton, OH   44720

Washington Technical College
Route 2
Marietta, OH   45750

Wilberforce University
Wilberforce, OH   45384

Wilmington College
Wilmington, OH   45177

OKLAHOMA
Bacone College
Muskogee, OK   74403

Bartlesville Wesleyan
College
2201 Silver-Lake Road
Bartlesville, OK   74006-6299

Bethany Nazarene College
6729 NW 39 Expressway
Bethany, OK   73008

Carl Albert Junior College
P.O. Box 606
Poteau, OK   74953

Connors State College
College Road
Warner, OK   74469

East Central University
Ada, OK   74820-6899

El Reno Junior College
P.O. Box 370
El Reno, OK   73036

Northern Oklahoma College
1220 East Grand Avenue
Tonkawa, OK   74653

Northwestern Oklahoma State
University
Alva, OK   73717

Oklahoma City Community
College
7777 South May Avenue
Oklahoma City, OK   73159

Oklahoma City University
NW 23rd and Blackwelder
Oklahoma City, OK   73106

Oklahoma State University
Technical Institute
900 North Portland
Oklahoma City, OK   73107

Oral Roberts University
7777 South Lewis
Tulsa, OK   74171

Phillips University
University Station
Enid, OK   73702

Rose State College
6420 SE 15th Street
Midwest City, OK   73110

Southeastern Oklahoma State
University
Durant, OK   74701

Opportunities for Credit

OKLAHOMA (cont.)
Southwestern Oklahoma State
 University
Weatherford, OK 73096

University of Oklahoma
660 Parrington Oval
Norman, OK 73019

University of Science and
 Arts of Oklahoma
Chickasha, OK 73018

Western Oklahoma State
 College
2801 North Main Street
Altus, OK 73521

OREGON
Bassist College
2000 SW Fifth Avenue
Portland, OR 97201

Blue Mountain Community
 College
P.O. Box 100
Pendleton, OR 97801-0100

Central Oregon Community
 College
College Way
Bend, OR 97701

Chemeketa Community College
P.O. Box 14007
Salem, OR 97309-5008

Clackamas Community College
19600 Molalla Avenue
Oregon City, OR 97045

Clatsop Community College
16th and Jerome
Astoria, OR 97103

Eastern Oregon State College
La Grande, OR 97850

Judson Baptist College
400 East Scenic Drive
The Dalles, OR 97058

Lane Community College
4000 East 30th Avenue
Eugene, OR 97405

Lewis and Clark College
0615 Southwest Palatine Hill
Portland, OR 97219

Linfield College
McMinnville, OR 97128-6894

Marylhurst College for
 Lifelong Learning
Marylhurst, OR 97036-0261

Mount Hood Community
 College
26000 Southeast Stark
Gresham, OR 97030

Portland Community College
12000 Southwest 49th Avenue
Portland, OR 97219

Portland State University
P.O. Box 751
Portland, OR 97207

Umpqua Community College
P.O Box 967
Roseburg, OR 97470-0226

University of Portland
5000 N Willamette Boulevard
Portland, OR 97203-5798

Warner Pacific College
2219 Southeast 68th Avenue
Portland, OR 97215-4099

Western Conservative Baptist
 Seminary
5511 SE Hawthorne Boulevard
Portland, OR 97215

PENNNSYLVANIA
Alliance College
Fullerton Avenue
Cambridge Springs, PA 16403

Alvernia College
Reading, PA 19607

Opportunities for Credit

Baptist Bible College of
 Pennsylvania
538 Venard Road
Clarks Summit, PA   18411

Bucks County Community
 College
Swamp Road
Newtown, PA   18940

Butler County Community
 College
College Drive Oak Hills
Butler, PA   16001

Cabrini College
Eagle-King of Prussia Roads
Radnor, PA   19087

Chatham College
Woodland Road
Pittsburgh, PA   15232

Chestnut Hill College
Chestnut Hill
Philadelphia, PA   19118-2695

Clarion University of
 Pennsylvania, Clarion
Clarion, PA   16214

Clarion University of
 Pennsylvania, Venango
1801 West First Street
Oil City, PA   16301-3297

Community College of
 Allegheny County, Allegheny
 Campus
808 Ridge Avenue
Pittsburgh, PA   15212

Community College of
 Allegheny County, South
 Campus
1750 Clairton Road
West Mifflin, PA   15122

Curtis Institute of Music
1726 Locust Street
Philadelphia, PA   19103

Delaware County Community
 College
Route 252 and Media Line
 Road
Media, PA   19063

Delaware Valley College of
Science and Agriculture
Doylestown, PA   18901

Eastern College
Saint Davids, PA   19087

Elizabethtown College
Elizabethtown, PA   17022

Geneva College
College Avenue
Beaver Falls, PA   15010

Gettysburg College
Gettysburg, PA   17325-1486

Gratz College
10th Street and Tabor Road
Philadelphia, PA   19141

Grove City College
Grove City, PA   16127

Gwynedd-Mercy College
Sumneytown Pike
Gwynedd Valley, PA   19437

Harcum Junior College
Bryn Mawr, PA   19010-3476

Harrisburg Area Community
 College
3300 Cameron Street Road
Harrisburg, PA   17110-2999

Indiana University of
 Pennsylvania
Indiana, PA   15705

Lackawanna Junior College
901 Prospect Avenue
Scranton, PA   18505

La Roche College
9000 Babcock Boulevard
Pittsburgh, PA   15237-5828

Opportunities for Credit

PENNSYLVANIA (cont.)
Lebanon Valley College
Annville, PA   17003-0501

Lutheran Theological
Seminary, Gettysburg
61 West Confederate Avenue
Gettysburg, PA   17325

Lycoming College
Williamsport, PA   17701-5192

Mansfield University of
Pennsylvania
Mansfield, PA   16933

Marywood College
2300 Adams Avenue
Scranton, PA   18509-1598

Mercyhurst College
501 East 38th Street
Erie, PA   16546

Millersville University of
Pennsylvania
Millersville, PA   17551

Moravian College
Bethlehem, PA   18018

Mount Aloysius Junior
College
Cresson, PA   16630

Neumann College
Aston, PA   19014

Northeastern Christian
Junior College
1860 Montgomery Avenue
Villanova, PA   19085

Penn State, DuBois
College Place
Dubois, PA   15801

Point Park College
201 Wood Street
Pittsburgh, PA   15222

Reading Area Community
College
10 South Second Street
P.O. Box 1706
Reading, PA   19603

Saint Joseph's University
5600 City Avenue
Philadelphia, PA   19131

Seton Hill College
Greensburg, PA   15601

Slippery Rock University
Slippery Rock, PA
16057-9989

Spring Garden College
102 East Mermaid Lane
Chestnut Hill, PA   19119

Susquehanna University
University Avenue
Selinsgrove, PA   17870-9989

Temple University
Philadelphia, PA   19122

University of Pittsburgh,
Pittsburgh
Pittsburgh, PA   15260-0001

University of Pittsburgh,
Titusville
504 East Main Street
Titusville, PA   16354-2097

University of Scranton
Scranton, PA   18510

Valley Forge Christian
College
Charlestown Road
Phoenixville, PA   19460

Waynesburg College
51 West College Street
Waynesburg, PA   15370

West Chester University
University and High Street
West Chester, PA   19383

Opportunities for Credit

Westmoreland County
 Community College
Youngwood, PA   15697-1895

Wilkes College
Box 111, 170 South Franklin
 South
Wilkes-Barre, PA   18702

Williamsport Area Community
 College
1005 West Third Street
Williamsport, PA   17701

Wilson College
Chambersburg, PA   17201

York College
Country Club Road
York, PA   17403-3426

RHODE ISLAND
Community College of Rhode
 Island
400 East Avenue
Warwick, RI   02886

Johnson and Wales College
Abbott Park Place
Providence, RI   02903-3776

Roger Williams College
Old Ferry Road
Bristol, RI   02809

Salve Regina College
Ochre Point Avenue
Newport, RI   02840

SOUTH CAROLINA
Aiken Technical College
P.O. Drawer 696
Aiken, SC   29802-0696

Chesterfield-Marlboro
 Technical College
Drawer 1007
Cheraw, SC   29520-1007

The Citadel
Charleston, SC   29409

Coker College
Hartsville, SC   29550

College of Charleston
Charleston, SC   29424

Columbia Bible College
P.O. Box 3122
Columbia, SC   29230-3122

Columbia College
Columbia College Drive
Columbia, SC   29203

Converse College
580 East Main Street
Spartanburg, SC   29301

Horry-Georgetown Technical
 College
P.O. Box 1966
Conway, SC   29526

Lutheran Theological
 Southern   Seminary
4201 North Main Street
Columbia, SC   29203

Midlands Technical College
P.O. Box 2408
Columbia, SC   29202

Newberry College
2100 College
Newberry, SC   29108

South Carolina State College
Orangeburg, SC   29117

University of South
 Carolina, Aiken
171 University Parkway
Aiken, SC   29801

University of South
 Carolina, Coastal
 Carolina College
P.O. Box 1954
Conway, SC   29526

University of South
 Carolina, Union
P.O. Drawer 729
Union, SC   29379

Opportunities for Credit

SOUTH CAROLINA (cont.)
Williamsburg Technical
   College
601 Lane Road
Kingstree, SC   29556

Winthrop College
Oakland Avenue
Rock Hill, SC   29733

SOUTH DAKOTA
Augustana College
29th and Summit
Sioux Falls, SD   57197

Black Hills State College
1200 University Street
Spearfish, SD   57783

Dakota State College
Madison, SD   57042

Mount Marty College
1105 West 8th Street
Yankton, SD   57078

National College
321 Kansas City Street
Rapid City, SD   57701

Presentation College
1500 North Main
Aberdeen, SD   57401-1299

Sioux Falls College
1501 South Prairie Avenue
Sioux Falls, SD   57105-1699

TENNESSEE
Austin Peay State University
College Street
Clarksville, TN   37040

Belmont College
1800 Belmont Boulevard
Nashville, TN   37203

Bethel College
Cherry Street
McKenzie, TN   38201

Bristol College
P.O. Box 757
Bristol, TN   37621-0757

Christian Brothers College
650 East Parkway South
Memphis, TN   38104

Cleveland State Community
   College
P.O. Box 3570
Cleveland, TN   37320-3570

Columbia State Community
   College
Hampshire Pike
Columbia, TN   38402-1315

Cumberland University
Lebanon, TN   37087

East Tennessee State
   University
Johnson City, TN   37614-0002

Fisk University
17th Avenue North
Nashville, TN   37203

Free Will Baptist Bible
   College
3606 West End Avenue
Nashville, TN   37205-2498

King College
Bristol, TN   37620

Memphis Academy of Arts
Overton Park
Memphis, TN   38112

Memphis State University
Memphis, TN   38152

Mid-South Bible College
P.O. Box 12144
Memphis, TN   38182-0144

Nashville State Technical
   Institute
120 White Bridge Road
Nashville, TN   37209

Roane State Community
   College
Harriman, TN   37748

Opportunities for Credit

Shelby State Community
College
P.O. Box 40568
Memphis, TN 38174-0568

Southern College of
Seventh-Day Adventists
Box 370
Collegedale, TN 37315-0370

State Technical Institute,
Memphis
5983 Macon Cove
Memphis, TN 38134

Tennessee Technological
University
Cookeville, TN 38505

Trevecca Nazarene College
333 Murfreesboro Road
Nashville, TN 37203-4411

Tusculum College
P.O. Box 5049
Greeneville, TN 37743-9997

University of the South
Sewanee, TN 37375-4013

University of Tennessee,
Knoxville
Knoxville, TN 37996-0154

University of Tennessee,
Martin
Martin, TN 38238-5009

Volunteer State Community
College
Nashville Pike
Gallatin, TN 37066-3188

TEXAS
Abilene Christian University
Abilene, Texas 79699

Amarillo College
P.O. Box 447
Amarillo, TX 79178

Amber University
1700 Eastgate Drive
Garland, TX 75041

American Technological
University
P.O. Box 1416 US Hwy 190
West
Killeen, TX 76540-1416

Austin College
900 North Grand
Sherman, TX 75090

Central Texas College
Highway 190 West
Killeen, TX 76541

Dallas Baptist University
7777 West Kiest Boulevard
Dallas, TX 75211-9800

Eastfield College
3737 Motley Drive
Mesquite, TX 75150-2099

East Texas Baptist
University
1209 North Grove Street
Marshall, TX 75670

East Texas State University,
Commerce
East Texas Station
Commerce, TX 75428

East Texas State University,
Texarkana
P.O. Box 5518
Texarkana, TX 75501

El Paso Community College
P.O Box 20500
El Paso, TX 79998

Hardin-Simmons University
2200 Hickory
Abilene, TX 79698

Hill Junior College
P.O. Box 619
Hillsboro, TX 76645

Houston Baptist University
7502 Fondren Road
Houston, TX 77074

Opportunities for Credit

TEXAS (cont.)
Howard County Junior College
1001 South Birdwell Lane
Big Spring, TX 79720-3799

Jacksonville College
Pine Street
Jacksonville, TX 75766

Lee College
P.O. Box 818
Baytown, TX 77522-0818

LeTourneau College
P.O. Box 7001
Longview, TX 75607

Lubbock Christian College
5601 West 19th
Lubbock, TX 79407

McLennan Community College
1400 College Drive
Waco, TX 76708

McMurry College
Sayles Boulevard & 14th
 Street
Abilene, TX 79697

Midwestern State University
3400 Taft Boulevard
Wichita Falls, TX 76308

Mountain View College
4849 West Illinois
Dallas, TX 75211-6599

Our Lady of the Lake
 University
411 SW 24th Street
San Antonio, TX 78285-0001

Panola Junior College
West Panola Street
Carthage, TX 75633

Saint Mary's University
One Camino Santa Maria
San Antonio, TX 78284-0400

San Jacinto College, Central
8060 Spencer Highway
Pasadena, TX 77505

Southwestern Adventist
 College
Keene, TX 76059

Southwestern Assemblies of
 God College
1200 Sycamore
Waxahachie, TX 75165

Southwestern University
University Avenue
Georgetown, TX 78626

Stephen F. Austin State
 University
1936 North Street
Nacogdoches, TX 75962

Texas A & I University
Santa Gertrudis
Kingsville, TX 78363

University of Dallas
University of Dallas Station
Irving, Texas 75061

University of Houston,
University Park
4800 Calhoun
Houston, TX 77004

University of Mary
 Hardin-Baylor
M H-B Station
Belton, TX 76513

University of Saint Thomas
3812 Montrose Boulevard
Houston, TX 77006

University of Texas, Dallas
P.O. Box 830688
Richardson, TX 75083-0688

Vernon Regional Junior
 College
4400 College Drive
Vernon, TX 76384

Weatherford College
308 East Park Avenue
Weatherford, TX 76086

Your Hidden Credentials

Opportunities for Credit

Wharton County Junior College
911 Boling Highway
Wharton, TX   77488

UTAH
College of Eastern Utah
400 East 4th North
Price, UT   84501

Latter Day Saints Business
 College
411 East South Temple
Salt Lake City, UT   84111

Southern Utah State College
351 West Center
Cedar City, UT   84720

Utah Technical College,
 Provo
P.O. Box 1609
Provo, UT   84603

Weber State College
3950 Harrison Boulevard
Ogden, UT   84408-1011

Westminster College of Salt
 Lake City
1840 South 13th East
Salt Lake City, UT   84105

VERMONT
Burlington College
95 North Avenue
Burlington, VT   05401-8477

Castleton State College
Castleton, VT   05735

Goddard College
Plainfield, VT   05667

Johnson State College
Johnson, VT   05656

Lyndon State College
Lyndonville, VT   05851

Norwich University
Northfield, VT   05663

School for International
 Training
Kipling Road
Brattleboro, VT   05301

Southern Vermont College
Monument Road
Bennington, VT   05201

Trinity College
Colchester Avenue
Burlington, VT   05401

Vermont Technical College
Randolph Center, VT   05061

VIRGINIA
Averett College
420 West Main Street
Danville, VA   24541

Blue Ridge Community College
P.O. Box 80
Weyers Cave,   VA 24486

Christopher Newport College
50 Shoe Lane
Newport News, VA   23606

Ferrum College
Ferrum, VA   24088-9001

George Mason University
4400 University Drive
Fairfax, VA   22030

Germanna Community College
P.O Box 339
Locust Grove, VA   22508

Hampton University
Hampton, VA   23668

James Madison University
Harrisonburg, VA   22807

John Tyler Community College
Drawer T
Chester, VA   23831-5399

Longwood College
Farmville, VA   23901

Opportunities for Credit

VIRGINIA (cont.)
Mary Baldwin College
Staunton, VA 24401

Mary Washington College
Fredericksburg, VA
22401-5358

Marymount College of Virginia
2807 North Glebe Road
Arlington, VA 22207

Norfolk State University
2401 Corprew Avenue
Norfolk, VA 23504

Northern Virginia Community
 College
4001 Wakefield Chapel Road
Annandale, VA 22003

Randolph-Macon College
Ashland, VA 23005-1698

Richard Bland College of the
 College of William and Mary
Route 1 Box 77-A
Petersburg, VA 23805

Southern Seminary Junior
 College
Buena Vista, VA 24416

Southwest Virginia Community
 College
Box S V C C
Richlands, VA 24641

Tidewater Community College
State Route 135
Portsmouth, VA 23703

University of Richmond
Richmond, VA 23173-1903

Virginia Commonwealth
 University
910 West Franklin Street
Richmond, VA 23284-0001

Virginia Highlands Community
 College
P.O. Box 828
Abingdon, VA 24210

Virginia Intermont College
Harmling Street
Bristol, VA 24201

Virginia State University
Petersburg, VA 23803

Virginia Wesleyan College
Wesleyan Drive
Norfolk, VA 23502-5599

WASHINGTON
Bellevue Community College
3000 Landerholm Circle SE
Bellevue, WA 98009-2037

Big Bend Community College
Andrews and 24th
Moses Lake, WA 98837

Clark College
1800 East McLoughlin
 Boulevard
Vancouver, WA 98663

Cornish Institute
710 East Roy
Seattle, WA 98102-4696

Eastern Washington University
Cheney, WA 99004

Grays Harbor College
Aberdeen, WA 98520-7599

Green River Community College
12401 South East 320
Auburn, WA 98002

Highline Community College
240th and Pacific Highway S
Midway, WA 98032-0424

Lower Columbia College
Longview, WA 98632

North Seattle Community
 College
9600 College Way North
Seattle, WA 98103

Olympic College
16th & Chester
Bremertown, WA 98310-1699

Your Hidden Credentials

Opportunities for Credit

Pacific Lutheran University
Tacoma, WA   98447-003

Saint Martin's College
700 College Street Southeast
Lacey, WA   98503

Seattle Central Community
 College
1701 Broadway
Seattle, WA   98122

Skagit Valley College
2405 College Way
Mount Vernon, WA   98273

Tacoma Community College
5900 South 12th Street
Tacoma, WA   98465

Walla Walla Community College
500 Tausick Way
Walla Walla, WA   99362

Washington State University
Pullman, WA   99164-1036

Whatcom Community College
5217 Northwest Road
Bellingham, WA   98226

Yakima Valley Community
 College
16th & Nob Hill Boulevard
Yakima, WA   98902

WEST VIRGINIA
Concord College
Athens, WV   24712

Fairmont State College
Locust Avenue
Fairmont, WV   26554

Glenville State College
200 High Street
Glenville, WV   26351-9990

Marshall University
Huntington, WV   25701

Shepherd College
Shepherdstown, WV   25443

Southern West Virginia
 Community College
Box 2900
Logan, WV   25601-2900

University of Charleston
2300 MacCorkle Avenue SE
Charleston, WV   25304-1099

West Liberty State College
West Liberty, WV   26074

West Virginia Northern
 Community College
College Square
Wheeling, WV   26003

West Virginia State College
Institute, WV   25112

West Virginia University
Morgantown, WV   26506

West Virginia Wesleyan
 College
Buckhannon, WV   26201

Wheeling College
Wheeling, WV   26003

WISCONSIN
Beloit College
Beloit, WI   53511

Cardinal Stritch College
6801 North Yates Road
Milwaukee, WI   53217

Carroll College
100 North East Avenue
Waukesha, WI   53186

Gateway Technical Institute
1001 Main Street
Racine, WI   53403

Lakeland College
P.O Box 359
Sheboygan, WI   53081

Marian College
45 South National Avenue
Fond du Lac, WI   54935-4699

Opportunities for Credit

WISCONSIN (cont.)
Milwaukee Area Technical
  College
1015 North 6th Street
Milwaukee, WI 53203

Mount Mary College
2900 Menomonee River Parkway
Milwaukee, WI 53222

Mount Senario College
College Avenue West
Ladysmith, WI 54848

Nicolet College and Technical
  Institute
Box 518
Rhinelander, WI 54501

North Central Technical
  Institute
1000 Campus Drive
Wausau, WI 54401

Northeast Wisconsin Technical
  Institute
2740 West Mason Street
P.O. Box 19042
Green Bay, WI 54307-9042

Northland College
1411 Ellis Avenue
Ashland, WI 54806-3999

Saint Francis Seminary,
  School of Pastoral Ministry
3257 South Lake Drive
Milwaukee, WI 53207-0905

Saint Norbert College
De Pere, WI 54115

Silver Lake College
2406 South Alverno Road
Manitowoc, WI 54220-9319

University of Wisconsin,
  Green Bay
Green Bay, WI 54301-7001

University of Wisconsin,
  LaCrosse
1725 State Street
LaCrosse, WI 54601

University of Wisconsin,
  Madison
500 Lincoln Drive
Madison, WI 53706

University of Wisconsin,
  Milwaukee
P.O. Box 413
Milwaukee, WI 53201

University of Wisconsin,
  Platteville
725 West Main Street
Platteville, WI 53818-9998

University of Wisconsin,
  Stevens Point
Stevens Point, WI 54481

University of Wisconsin,
  Superior
1800 Grand Avenue
Superior, WI 54880-2898

Viterbo College
815 South 9th Street
LaCrosse, WI 54601

Waukesha County Technical
  Institute
800 Main Street
Pewaukee, WI 53072

Western Wisconsin Technical
  Institute
6th and Vine Streets
LaCrosse, WI 54601

Wisconsin Indianhead
  Technical Institute
P.O. Box B
Shell Lake, WI 54871

WYOMING
Casper College
125 College Drive
Casper, WY 82601

Central Wyoming College
2660 Peck Avenue
Riverton, WY 82501-1520

Opportunities for Credit

Laramie County Community
 College
1400   East   College   Drive
Cheyenne, WY   82007-3299

University of Wyoming
Box 3434 University Station
Laramie, WY   82071

PUERTO RICO
  Antillian College
  Box 118
  Mayaguez, PR   00709-0118

Humacao University College
CUH Station
Humacao, PR   00661

Turabo University
P.O. Box 1091
Caguas, PR   00625

QUESTIONS	Goddard College Off-Campus Study Plainfield, VT 05667	Empire State College Center For Distance Learning 2 Union Avenue Saratoga Springs, NY 12866
What degrees can I earn in the program? Some independent learning colleges award only one degree. Others offer several.	Bachelor's and Master's degrees	Associate, Bachelor's and Master's degrees
What can I study in the program? Each independent learning program offers degrees in specific areas. Some also offer "general" or "interdisciplinary" degrees that allow you to study a wide range of topics from many disciplines.	Leadership and Management, Education, Psychology, Environmental Sciences, and Interdisciplinary Creative and Liberal Arts	Business, Human Services, Interdisciplinary Studies
What are the requirements for entrance into the program? Some independent learning programs require a high school diploma or equivalency; others do not. In addition, because of the special challenge of independent study, some programs require evidence of the ability to study on your own.	High School Diploma or Equivalency; Evidence of commitment to and basic ability for self-direction.	High School Diploma or Equivalency
Can I study part-time? Some programs allow part-time study. Others do not.	Full-time Study Only	Full- or Part-time Study
How long does each semester last? From 12 to 24 weeks, depending on the program. Some programs allow you to study at your own pace, with no semesters at all.	20-week semesters; 15 semester hours of advancement	16-week semesters
How do I get credit for my previous learning? Like many college programs for adults, independent learning colleges award credit for previous learning in a number of ways: transfer credit from other colleges; national standardized exams; apprentice, military and other training; and individual "life experience" portfolios.	Transfer Credits, Portfolios, Assessment, and Accepted Exams for Bachelor's degrees; Bachelor's degree required for M.A. study.	Transfer Credits, Exams, Military Education, and Approved Training Programs
How much of my degree can I earn through my previous learning? Each program is different, and all credits are subject to university requirements. Please keep in mind that very few students earn the maximum amounts listed here.	Up to 25% in portfolio assessment and up to 75% in transfer credit, but no more than 75% of a bachelor's degree in all	Up to 62% of an Associate or 75% of a Bachelor's degree.
How are independent study "courses" put together? Some programs have pre-set, carefully structured learning materials designed for adult independent study. In other programs, you and an instructor design an individualized study "contract" or plan just for you.	Individual Study Plans	Pre-set Materials
Independent study sounds lonely. How will I keep in touch with my professors and my fellow students? There's no doubt—these programs are for people with the discipline to study on their own. But each program provides ways to keep you in touch with counselors and instructors and, often, your fellow students.	Each semester begins with a 9-day stay at the Goddard campus in Vermont. There you meet with your advisor and make contact with other students. For the rest of the semester, you correspond regularly with your advisor.	Students work through their course materials under the guidance of a faculty member maintaining regular contact by phone or mail.
How much does each program cost? Each is different, and much depends on whether you are studying full- or part-time. REMEMBER: Union and company tuition benefits can help a great deal with tuition costs.	$2150 per semester for undergraduate work; $2400 per semester for Master's study	$57.00 per credit. (Most courses are 4 or 8 credits.)

Provided courtesy of CAEL (Council for Adult and Experiential Learning)

Your Hidden Credentials

American Open University of New York Institute of Technology Central Islip, NY 11722	The Union for Experimenting Colleges and Universities Undergraduate Studies 632 Vine Street Suite 1010 Cincinnati, OH 45202	Regents College of the University of the State of New York Cultural Education Center Albany, NY 12230	Ohio University External Student Program 309 Tupper Hall Athens, OH 45701	Thomas A. Edison State College 101 West State Street Trenton, NJ 08625
Bachelor's degree	All Degrees from Bachelor's through Doctoral	Associate and Bachelor's degrees	Associate and Bachelor's degrees	Associate and Bachelor's degrees
Business Administration, General Studies and Behavioral Sciences, with options in Mental Health, Psychology, Sociology, and Criminal Justice	Business, Behavioral Sciences, Human Services, Health Care, and a wide range of Liberal Arts	Computers, Business, Electronics, Nuclear Technology, Nursing, and a wide range of Liberal Arts and Sciences	Associate degrees in Security and Safety Technology and a wide range of Liberal Arts and Sciences; Bachelor's degree in General Studies	Business and Management, The Liberal Arts, Applied Science and Technology, Human and Social Services, Nursing
High School Diploma or Equivalency	High School Diploma or Equivalency; Evidence of motivation and sense of educational responsibility	No Entrance Requirements; Students must be self-directed.	High School Diploma or Equivalency	No Entrance Requirements
Full- or Part-time Study	Full-time Only	Full- or Part-time Study	Full- or Part-time Study	Full- or Part-time Study
You have up to 6 months to complete any course but can register for new courses at any time.	13-week quarters	No Set Semesters	No Set Semesters	No Set Semesters
Transfer Credits, Exams, Military Education, Approved Training Programs, and Portfolio Assessment	Transfer Credits, Exams, Military Education, Approved Training Programs, and Portfolio Assessment	Transfer Credits, Exams, Military Education, Approved Training Programs, and Special Assessment	Transfer Credits, Exams, Military Education, Approved Training Programs, and Portfolio Assessment	Transfer Credits, Exams, Military Education, Approved Training Programs, Media-assisted instruction, Portfolio Assessment
Up to 90% of a Bachelor's degree	Up to 75% of a Bachelor's degree	No Limit	Up to 100% of an Associate or Bachelor's degree through portfolio assessment and other appropriate credit	No Limit
Pre-set Materials	Individualized Learning Contracts	Varied Methods	Pre-set Materials	Varied Methods
Students correspond with their professors by mail. In addition, a special computer conferencing system enables students to "talk" with their professors and fellow students.	In addition to its central faculty in Cincinnati, UECU has faculty members in your region with whom you meet regularly to plan and discuss your work.	Regents College has no classes. By mail or phone, your advisor helps you identify educational resources that, together with your previous learning, enable you to earn a degree. The Graduate Resource Network gives you local support and help.	Correspondence courses include a textbook and study guide. The lessons you do are submitted to OU faculty for grading and feedback. In addition, midterm and final exams are taken with OU proctors in your local area.	Edison College has no classes. You are encouraged to stay in touch with advisors by mail or phone. An advisement center is available. Nursing Program has Peer Study Groups. Staff and workshops are available for special college programs.
$50 application fee; $100 matriculation fee; $75 per credit	$1300 per quarter	$225 enrollment fee; $125-175 per year records fee; $60 graduation fee; you pay for courses and exams directly	$50 admission fee; $35 matriculation fee per year; $31 per quarter hour for correspondence course and $17 per quarter hour for exam credit; $5 enrollment fee per course	$50 application fee; $125 (NJ) - $200 (others) First Year Tuition Equivalency Fee; $100 (NJ) - $150 (others) Subsequent Year's Tuition Equivalency Fee; $70 Graduation Fee

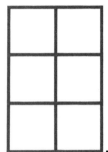

Bibliography

1. Aslanian, et al. *40 Million Americans in Career Transition: The Need For Information.* New York: Future Directions for a Learning Society, The College Board, 1978.

2. Baum, L. Frank. *The Wizard of Oz.* Scholastic Book Services. New York, 1958.

3. Cross, Patricia K. *Adults as Learners: Increasing Participation and Facilitating Learning.* San Francisco: Jossey-Bass, Inc. 1981.

4. Dewey, John. *Experience and Education.* Collier Books. Collier-MacMillan: London, 1938.

5. Gross, Ronald. *The Lifelong Learner: A Guide to Self-Development.* New York: Simon and Schuster, Inc. 1977.

6. Knowles, Malcolm. *The Adult Learner.* Houston: Gulf Publishing Company, Book Division, 1984.

7. Tough, Allen. *The Adult's Learning Projects.* Toronto: The Ontario Institute for Studies in Education, 1979.

8. Tough, Allen. *Intentional Changes: A Fresh Approach to Helping People Change.* Chicago: Follett Publishing Company, 1982.

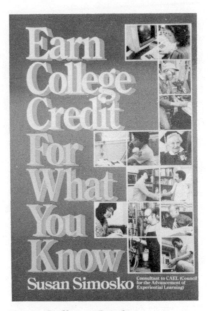

Earn College Credit For What You Know

by Susan Simosko
Consultant to CAEL
(Council for Adult and
Experiential Learning)

Now you can earn credit for college-level courses—without entering a classroom!

All you have to do is document the outside knowledge you've gained through your job, community service, hobbies, reading, special interests, and accomplishments. And *Earn College Credit for What You Know* will show you how to put together your "portfolio of experience" to gather appropriate background information and document it for your chosen accredited college from among 571 listed by name and address in the book that have prior learning assessment programs.

ISBN 87491-773-5/$8.95 quality paper/
Study Aids
200 pages, 6 × 9, illustrations, index